YOU CAN'T DREAM BIG ENOUGH

You Can't Dream Big Enough

by

Orion Samuelson

The American Farmer's Best Friend
for Over 60 Years

with

Steve Alexander, Diane Montiel
and
Gloria Samuelson

Foreword by Mike Johanns

An imprint of Bantry Bay Media, LLC Chicago

Printed in the United States of America
at Lake Book Mfg., Melrose Park, Illinois

For information, address Bantry Bay Publishing at:
bantrybaymedia@gmail.com
Please visit our website: www.bantrybaybooks.com

ISBN 978-0-9850673-1-1

To schedule Orion Samuelson
for a speaking engagement or appearance,
please send an inquiry to: agspeakers@gmail.com

DISCLAIMER

While all care has been taken in the preparation of this memoir, no responsibility is accepted by the author, contributors or publisher, its staff or volunteers for any errors, omissions or inaccuracies. Sixty-plus years is a long time and if a story, name or location contains a mistake, you have our apologies and assurance that no malice was intended. Regarding photos, many of which date back to the 1950s, every attempt was made by the publisher and copyright holder to determine and assign proper credit and obtain permission for the use of photos, but some were impossible to trace. We gratefully acknowledge all who have given permission by tagging each photo accordingly. In general, no responsibility can be accepted by the author(s) or publisher for loss occurring to any person doing anything as a result of the material in this publication.

This book is dedicated to every member of my family, from grandparents to grandchildren, who supported and encouraged me to "dream big" and fulfill those dreams.

CONTENTS

YOU CAN'T DREAM BIG ENOUGH

Credit: 123RF.com

Introduction

For at least the last ten years, I have been talking about — and people have been talking *to* me about — writing a book. The biggest obstacle was that I am not a writer. I do my writing with my mouth. But there were other people who listened, wrote and edited, and you are holding that joint effort in your hands.

Another reason I put it off was that I didn't want any part of my book to seem as if I was bragging or calling attention to myself. Although I certainly don't shy away from the spotlight or podium if given the chance to talk about the accomplishments of the American farmer, my veins run full of Norwegian modesty. As a group, we Norskis tend to be a bit reserved, and some might argue that's an understatement.

But I wanted to do a book for my two grandchildren, Matthew and Grace, so that they, and maybe their children, someday, will have some sense of what I did. Being a farm broadcaster opened up my world in ways I never could have imagined. Traveling to 43 countries, having dinner at the White House, meeting seven presidents and witnessing first hand the highlights and low lights that our farmers and ranchers have faced over the years have created the memories and opinions that fill these pages. For a title, I had considered *The Manure Tour: Life with an Agricultural Broadcaster.* But the more thought I gave to this wonderful life I am living, the more I saw that what has happened to me is the result of having big dreams. Thus, *You Can't Dream Big Enough*!

Watching young people grow and make their way through college, it struck me that the best advice I can give any of them is to dream big, and not to worry about dreaming too big, because it isn't possible. There is a world out there they can't begin to imagine. Many will work in a job or industry that hasn't even been invented yet, so I tell them to get prepared with education and have big dreams. That doesn't mean all of their dreams will come true, but why limit opportunities with small, safe dreams?

As one of the commencement speakers at the 2012 University of Illinois graduation ceremonies where, ten years earlier, I was bestowed with an Honorary Doctor of Humane Letters degree at age 67, I told the students that they were doing what I was doing; writing a book. All of us are writing the books of our lives each day. From the moment we arrive on this planet, the book begins. Chapters should include "Lessons I Have Learned." One of the biggest lessons in my book is, "Never evaluate a happening as it's happening. Give it some time." In the following pages, you'll read about a couple of life-changing challenges that, as they were happening, I wasn't able to see the possibilities. I could only see the negatives.

Another lesson: "Losing isn't the end of the world." You will lose. We all do at some point and in some way. In one of my first, most public losses which you'll read about, my FFA advisor came to the rescue and said "Let's analyze what went wrong and learn from it." We did, and I'm a much better broadcaster because of it.

Another chapter: "Find something you like to do so much you'll do it for nothing and then do it so well, you'll get paid for it." That's what I have done and I haven't worked a day in over 60 years of being an agri-business broadcaster, something which my late father, a calloused Wisconsin dairy farmer, was quick to point out on one of his rare visits to Chicago to see me doing my job.

The best part of my job has been people. People I've worked with, people I've talked with and the people in my audience. I learned early in life the best "collectible" is friends and over the years I've been extremely fortunate in developing friendships and relationships, sometimes in unusual ways and you never know where they will lead you. Whether it's a President of the United States, a corn and soybean farmer in Illinois or a cattle rancher in Montana, people make my work a pleasure. In my travels abroad and crisscrossing the United States, I've met the most interesting people and when I'm on the air, I think about each of them being on the other end of my microphone. I put a face on it. It could be the face of any of hundreds of people.

There are many more chapters of my life than I have room to write. I'm grateful for all of my experiences and I hope you'll enjoy the ones I've chosen.

Everything I am and have done is based on faith and family. My upbringing in a small country church has carried me through life and I am so blessed with my wife, Gloria, my son, David, his wife Carla and their two children (my grandchildren) Matthew and Grace, and my daughter, Kathryn. Thanks to them, my glass is always half-full, never half-empty.

My career has spanned some unbelievable advances in technology. For a farm boy who spent his first 14 years without electricity and a telephone, the Digital Age has brought, and continues to bring, changes I couldn't have dreamed of. Throw in the always changing societal norms regarding privacy, modesty and language and it really is a new world. Most of the changes have been for the better. But there have also been a few for the worse, and I fully realize my take on this is probably partly due to generational differences. I'm bothered by the language. I've always believed that if you can't make your point without using four-letter words, you're obviously not able to make your point. I also don't think we've improved the world with some of the formerly taboo topics that are now covered publicly on the air.

WGN has certainly changed since I started there in 1960. It has had to, in order to keep up with technological changes and changes in our audience and their listening habits. But change is often difficult and I haven't liked all of the changes, but I've always believed that if you don't own the radio station, it really is out of your hands. You only have two choices: go or stay. I've obviously stayed.

Not all of the changes in the radio industry have been good for the public, whose interest we are supposed to be serving. Consolidation of ownership really battered local radio. I've done small town radio and I know how important it is to the community that it has a radio station broadcasting everything from birth announcements to funeral announcements. Consolidation made that sort of community involvement all but disappear in places like Minot, North Dakota, where, at one time, one company owned eight radio stations. But we've seen how the investment bankers squeezed cash out of those mega-companies and left them in financial ruin. Slowly, local ownership is returning, and local radio is getting back to doing what it does best.

Throughout the book, you'll notice that I often say "we," as in "We had been working on ethanol for over 30 years…" I've always considered myself to be part of the farming community, even though I've been doing my field work from behind a microphone. So, when I say "we," it isn't from any sense of ego or false ownership, it's because I never stopped thinking like a farmer and feeling like a farmer.

Finally, I consider myself a great teller of Ole and Lena stories and since there are and were Oles and Lenas in my family — both in name and in character — I feel qualified to tell Ole and Lena stories. I'll scatter them throughout the book, beginning with the one below.

Ole and Lena went to the hospital so Lena could give birth to their first baby. As Ole paced in the waiting room, the doctor came out to inform him that he had some good news and some bad news about the birth. "The good news is that you have a normal baby boy. The bad news is that it is a Caesarian."

Ole started crying. "Vell," he said to the doctor, "I'm glat it is a helty baby, but I vas kinda hoping it vould be a Norvegian."

OF AGRICULTURE
WASHINGTON

Foreword

by

U.S. Senator Mike Johanns
U.S. Secretary of Agriculture, 2005-2007

The most recognized voice in rural America belongs to one of the most respected ag broadcasters in our nation and now the curtain is pulled back on Orion Samuelson's life story, which is both uplifting and inspiring. He worked hard to earn his iconic status in rural America and readers will delight in knowing how he did it. Even more delightful, they will learn that behind that booming voice is a truly good man whose passion for American agriculture runs as deeply as our country's roots in the land and its cultivation.

Orion is living proof that the American dream is within reach for any young person eager enough to strive for it. His story is especially inspiring to youth in rural America, where it's easy to feel far removed from famous figures. I know, because as a kid I never dreamed there was any road that led all the way from my family's farm in rural Iowa to Washington, DC. Orion's story helps to teach every child that there is no farm too distant or community too small to be the birthplace of an icon. For the millions of members of 4-H and FFA across the country, Orion's story is proof positive that the skills and confidence they are acquiring today can one day lead them to great accomplishments.

I remember thinking as a young man how cool it would someday be to meet the broadcasting legend known as "The Big O." That opportunity first came when I was Governor of Nebraska, attending Husker Harvest Days. I quickly learned that Orion's questions would reach far beyond the obvious or simple, actually grasping the intangible and always driving to the heart of what is best for agriculture. Few people have greater command of the issues affecting agriculture. Mastering a subject is foundational to becoming a strong advocate for it and Orion has accomplished both.

My respect for Orion has only deepened throughout my public service. I recall as Secretary of Agriculture, when preparing to embark upon an unprecedented nationwide effort to solicit advice from producers for the next farm bill, I knew exactly who I wanted by my side: Orion Samuelson. He agreed to moderate my first USDA Farm Bill Forum and his involvement sent a signal across rural America about the importance of our undertaking. Rural America took notice and submitted more than 4,000 comments. I also asked Orion to be among the first to report on the farm bill proposal when I unveiled it. The guarantee of a fair story was worth the penetrating questions I knew he would pose.

Our paths crossed many times over the years and one such meeting still brings a smile to my face. We were at the Farm Progress Show in blazing 100 degree heat taping an hour-long interview in the open sun. I don't know which one of us produced more sweat, Orion or me, but between us we could have irrigated a full acre. Yet, he never once complained about the heat that day. When one thinks of broadcasting, the image of a climate-controlled studio comes to mind. Not for Orion Samuelson. He travels the world to be where the agriculture story is developing, even when it's at the end of a dirt road, outside a small town, baking in the blazing sun.

Orion's story is the American dream, but not one of the silver platter sort. The challenges he faced would have brought down a lesser man. He's a farm boy at heart whose work ethic, determination and resiliency fueled his success, much like the farmers and ranchers whom he so respects.

From one former farm kid to another, way to go, my friend! You have long been an inspiration to me and this book invites others to know the great man behind the voice and face of American agriculture.

Mona, Sid and "Little O" in 1934.

Chapter One

Humble Beginnings

On a summer day after I finished the eighth grade, Dr. Walter Jones walked out of the X-ray room at St. Francis Hospital in La Crosse, Wisconsin toward where my parents and I were waiting, looked at me and said, "I wish there was another way I could tell you this, young man, but there isn't. You aren't going to walk for two years."

La Crosse was 32 miles northwest of our family's farm near Ontario. Most years, Mom, Dad, my sister Norma, who is five years younger than me, and I made it to the "big city" only twice a year: in December to do Christmas shopping and in the spring to buy clothes and supplies for the summer and fall. The trip to see Dr. Jones was in the summer after my eighth-grade year. I was excited about going into high school because I would finally be leaving O'Connell School, a one-room, eight-grade country school with about 40 students. We all sat together in the same room and as a first-grader doing your work, you would hear what second-, third-, fourth-, fifth-, sixth-, seventh- and eighth-graders were being taught. By the time I got to sixth grade, the teacher, Esther Garnett said, "Your test scores are high, there's no sense in you doing sixth grade. We'll just promote you to seventh." After graduating eighth grade, I was ready for Ontario High School, five or six miles away, but I was going to have to wait to join the 22 other kids in my class.

We had 30 milk cows, six sows, 200 chickens, one dog, 13 cats and, for a time, a team of horses. Our farm was 200 acres, but most of them were up and down hills with trees and rocks. Only 90 were tillable. It wasn't what you would call prime farming land. My grandfather had settled there after coming here from Norway in the late 1800s and when I visited Norway for the first time, I saw that farms there were on hills, even steeper hills than ours. Farmers would carve out a shelf on the hillside, put up the barn and the house and then farm down the hill. That was much of what we had on our farm. It provided a living for us; we always had food on the table and clothes on our back. There were no vacations for our family, which wasn't uncommon on small dairy farms, because the cows had to be milked twice a day, seven days a week. Dad didn't trust anyone else to milk those cows, so we never made it very far away from the farm.

In 1872, when Grandma Jenny was eight, she, an older brother and an adult uncle were sent to America because times were so tough, not only in Norway, but in all of Northern Europe. America was the land of opportunity, so parents sent their children away, knowing they might never see them again, to give them a chance at a better life.

Late spring blizzards are not unusual in Wisconsin and the one that blew through on March 31, 1934 coincided with my birth. Mona Karina Paulson Samuelson and Sidney Orlando Samuelson were about to become parents for the first time. Mona was in labor, but the storm kept the doctor from getting out to the farm to guide me into the world. So my great-aunt acted as midwife, did the delivery and did it well, because here I am.

My parents wanted to name me in honor of my grandfathers Ole Samuelson and Carl Paulson by choosing names beginning with O and C. They thought Orrin was too common, so they used a slight variation, Orion, which is not very common. I've met only three other Orions in my life. My name is misspelled or mispronounced or just missed altogether probably 70% of the time. I get called Orrin a lot. And there are those who call me "oh-RYE-uhn," as in the constellation. I've had people tell me, "Well, your parents knew you were going to be a star, so they named you after the constellation." Orion Clifford Samuelson was born in the bedroom of the farm home where I ultimately grew up; the same bedroom where I wound up bedridden most of my ninth and tenth grades.

I loved school and was excited about moving on to Ontario High School. I was a tall kid, so I was going to play basketball. The school wasn't big enough to have a football team, but we did have basketball and baseball. Early in the summer after eighth grade, I started experiencing pain in my left hip and it got worse rather quickly. It felt like something had come out of joint. My parents believed in chiropractic, so they took me to see three or four chiropractors, all of whom said they were going to cure me. But the pain just got worse. By August, they decided something else had to be done.

That morning before we drove to see Dr. Jones started like any other morning on the farm. At 5 o'clock, I helped Dad feed and milk the cows. We cleaned the barn and then got ourselves cleaned up for the trip to the hospital. We figured we would be back in time to do the evening milking. In those days, you had to wait awhile on X-rays. The doctor told us to go out and have dinner — which is what we called our noontime meal — and come back in the afternoon for the results.

After Dr. Jones dropped the bomb on us, I just stared at him in astonishment for a moment, letting what he had just said soak in. Then I asked, "What do you mean I'm not going to walk for two years?"

"I mean you aren't going to walk for two years. You have Legg-Perthes Disease," which he went on to explain is basically a decaying of the bone that goes from the thigh and fits into the hip socket. "Your leg is already shortening, which is what is causing the pain. If we don't treat it, you'll wind up with that leg three or four inches shorter than the other. I want to put you in the hospital right now."

That was the only time I saw my dad cry. The doctor left us to talk it over and once we regained our composure, we decided we'd better get going on this, we'd better do it. So, I was admitted to the hospital that day and was there for three weeks with a rope, pulley and 20-pound weight pulling on my leg. After that, I was in a cast. First, a body cast for six weeks. The first three weeks, I was in the hospital, but then the doctor let my parents take me home to my bed, where all I could do was lie in that itchy body cast in Wisconsin's summer heat and humidity. After the body cast, I was in a wheelchair for four months and finally crutches. It was a terrible time and I thought it was just the worst possible thing that could have happened to me. "God, why me?" I asked in frustration.

So I wasn't in the best of moods when the Ontario High School Vocational Agriculture teacher drove his car into our farmyard. Robert Gehring made a point of going to each of his incoming freshmen's farms to encourage them to take agriculture courses and join the Future Farmers of America.

"I'm here to welcome you to high school," he said with a smile.

I snapped at him, "You're wasting your time. I'm not going to be in school for two years."

He asked why and I told him the story. As we talked, I noticed that he had an artificial arm. He caught me looking at it and explained that when he was six, he lost his arm in a farm accident.

"Do you think you can study by yourself?" he asked.

"What do you mean?"

"Can you study without a teacher and students around you and keep up with your schoolwork?"

"Well, I don't know, but I think so."

"If you can give it a try, I'll make the five-mile trip out from town two or three times a week. I'll bring the assignments from the other teachers and I'll take your finished work to them so you don't fall behind. When you're back on your feet, you'll still be in the same grade as your friends and you can graduate on time."

Mr. Gehring's kindness was unexpected but I was very grateful because I'd made some real good friends in the eighth grade at O'Connell School and we were going into high school together. I wanted to keep up with them. Mr. Gehring was mild-mannered, almost meek at times, with a soft voice. It bothered me when I finally got to school and saw some students taking advantage of him.

When I wasn't doing schoolwork, I killed time by listening to the radio. This was before we had electricity on the farm, which I'll write about later. We had a radio that ran on "B" batteries, which were rationed during World War II. So, we had to be very careful about our radio use. But, by the time I was confined to my bed, the war had ended and the batteries were available again. I listened to Bert Wilson doing Cubs games on WIND. WLS was the Prairie Farmer Station back then and had the *National Barn Dance*, featuring performers who went on to become big stars, like Gene Autry, George Gobel, Pat Buttram and others.

The more I listened to the various announcers, the more I thought that maybe it was something I might be able to do because physically, I wasn't going to be able to do what I'd always figured I would: take over the family farm when Dad retired and be as good a dairy farmer as I could. The doctor warned that if I took a severe jolt, my condition could restart. Farming was all I knew, really. That's where my world ended. I didn't see much beyond that. You know, I'd go to La Crosse, which I think was a town of about 25,000 then, and I'd say, "Oh, that's way too big, too crowded. I don't want to live there!" But, lying in bed, listening to the radio, my world was opening up. Now, I was about as bashful and barefoot a country boy as you could find and the thought of me talking on the radio seemed so ridiculous, but I spoke to my Vo-Ag teacher about broadcasting. He told me that he'd get me into public speaking and when I was back on my feet, I would be in FFA speaking contests.

When I was strong enough to go back to school in my junior year, I started making speeches. I got into Forensics classes with Mrs. Woods, my English teacher, and by my senior year I was pretty good and was a state finalist in the Wisconsin FFA public speaking contest. Well, I went to that convention telling myself, "Man, I'm going to win this for my little high school." There were five of us. I came in fourth! I got my tail whipped. I was shocked. I asked Mr. Gehring, "Gee, what did I do wrong?" In his calm way, he told me that we would take some time, take a look at it and learn from it.

That's one lesson I took from my illness and those two lost years, a lesson I've used throughout my life: *Don't evaluate a happening when it's happening. Give it some time because God may have a reason.* What was a tragedy at the time changed my entire life, totally changed it. Had it not happened, I'd probably be milking cows in Wisconsin, and be happy doing it (except for the price of milk).

I might also have become a Lutheran preacher. We belonged to the Brush Creek Lutheran Church, where I was baptized and confirmed and where I spent a good deal of time. As is typical with small, country churches, it was the social and spiritual center of the community and a lot of our social life centered around the church. The Ladies Aid church dinners were held at noon on Sundays and there was an organization for kids called Luther League that we were involved in. I will say that the Ladies Aid dinners pretty much cured me of enjoying a modern-day buffet because I found myself standing in line every Sunday waiting to get something to eat. After growing up doing that, I decided I didn't need to do that as an adult. But, I do, because many

Brush Creek Lutheran Church, my home church, near Ontario, Wisconsin.

of the events and speaking engagements I attend have buffets.

Our little church didn't have enough members to support a full-time pastor, so we had a traveling minister. We only had church services every other week. On Christmas and Easter, the pastor, who was fluent in Norwegian, preached in the old country language. Until I was 11 or 12, I prayed the Lord's Prayer in Norwegian every night:

Fader vår, du som er i himmelen!
La ditt navn holdes hellig. La ditt rike komme. La din vilje skje på jorden som i himmelen.
Gi oss i dag vårt daglige brød.
Forlat oss vår skyld,
som vi og forlater våre skyldnere.
Led oss ikke inn i fristelse,
men frels oss fra det onde.
For riket er ditt, og makten og æren i evighet. Amen.

And I can still do the table prayer in Norwegian:

I Jesu navn går vi til bords,
å spise, drikke på ditt ord.

Deg, Gud til ære, oss til gavn,
Så får vi mat i Jesu navn. Amen.

When I was 24 or 25, my radio career already underway, I seriously considered becoming a pastor. But I had not gone to college, which was mandatory if I was to be a pastor, so that meant four years of college and four years of seminary, pushing me into my early thirties. I just didn't think that I would be able to get that done, so I settled with trying to be the best layman I could be and I have been as active as I can be in church. I was married in a Lutheran church in 1955 and my two children were adopted through Lutheran Social Services, another close, personal connection to my religion. But it was interesting, because if you were Norwegian, you were Lutheran and if you were German-Irish, you were Catholic, and in those days, never the twain shall meet. You stuck to your own and didn't cross over and date kids of the other faith. The one time I made my mother cry was when she found out I had taken a Catholic cheerleader home from a basketball game. Oh my, she was crushed. She had a vision of me marrying that girl and "raising 12 kids for the Pope!"

Our little church had a Christmas program each year and all of the young people had to get up in front of everybody and deliver their "piece," either a musical number or some sort of recitation. My mother and father gave me plenty of advance prompting to stand up tall and speak slowly, clearly and loud. Standing in front of the altar and facing the congregation, it seemed like a huge auditorium. Going back to that church as an adult, I had to laugh, because there are only ten rows of pews in that tiny sanctuary, but it sure didn't look that way to the eyes of a child.

My church experience growing up led to the most popular minute on our television show, starting in 1975 with the U.S. Farm Report and going for the next 30 years that we produced that program. Realizing how important Brush Creek Lutheran Church was to our family and community, we decided to do a one-minute Country Church Salute each week. I invited members of our audience to send photographs and a brief history of their country church. That feature prompted more reaction from our audience than anything else we did. My one regret is that we didn't save each of those histories and those photographs and publish a book, because I could have sold a book to every member of every congregation featured on the TV show.

My grandparents spoke Norwegian 90% of the time and I learned to understand it because I knew that when they and my parents were speaking Norwegian, it was often because they didn't want me to know what they were talking about. Of course, I wanted to know. I understood it much better than I was able to speak it, and I really can't read it at all. But when I got to be about 12, it wasn't cool if you didn't always speak English. Other kids would make fun of you, so you kind of put it aside.

NORMA

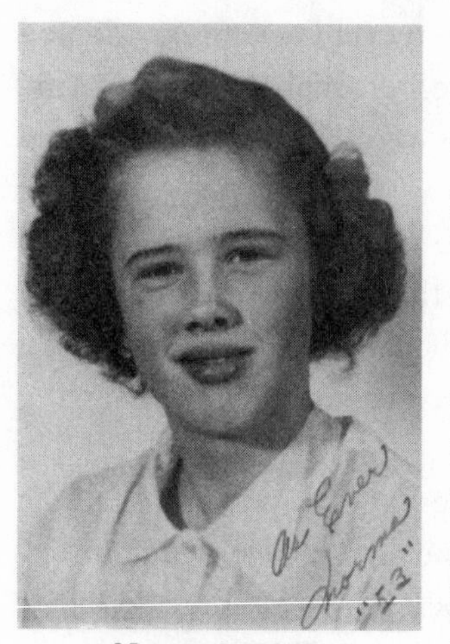

Norma in 1953.

My sister Norma arrived on the planet five years after I did and I suppose we had the typical brother-sister relationship. We did fight and go back and forth a lot. Of course, as we got older, we found that we liked each other. Rarely a day goes by that I don't think about what an unsung hero Norma is. While I get attention and recognition simply because I have a radio transmitter and a television network, Norma is the one who truly deserves the attention and recognition. She stayed in the area, married a local boy and over a 28-year period, Norma and Norman Haakenstad had 130 foster children go through their home. They had four children of their own and later adopted one of the foster children. Those 130 children were given a safe and loving place to live until the courts sent them back to their parents or to adoptive parents. Many of them were difficult children with behavioral problems and who had parents with addiction problems, but Norma and Norman gave them structure, discipline and love. How she was able to do that, I will never understand because I know I couldn't. Norma and Norman had been married 52 years when he passed away in January of 2012.

Secondly, Norma took care of our mother and father in their later years when they became bedridden. Our dad lived to be 95, our mother died at 89 and the peace of mind I received knowing that Norma was there for them, providing end of life care for them, well, it's something I can never repay her for. It's simply above and beyond the call of duty and I'll always be grateful to her and admire her for it.

One footnote about Mr. Gehring, who, outside of my parents, had more influence on my career than anyone. I lost track of him. I knew where the Ontario High School basketball coach was who let me do the P.A. announcing at basketball games, but I lost track of Mr. Gehring. In 1998, I asked Norma, if she could help me find him. She tracked him down in Milwaukee, where he had gone to live after retiring. Norma called him on the phone and said, "My brother Orion and I were students at Ontario High School." He said, "Oh, yeah, yeah, I remember you kids. So, what's Orion doing?" I loved it, I loved it! He had no idea that I was on radio and television, in large part because of him. I talked with him and filled him in on my career and had a nice conversation. Mr. Gehring changed my life and whenever I speak to FFA kids and other young people, I tell them to treat their Vo-Ag teachers with respect, because man, they do good, important work!

Back to my illness; the medical costs and inconvenience to my parents were substantial. There was no such thing as health insurance back then, at least for us. Dad didn't have health or life insurance. Plus, he lost a farm worker. I had gotten big and strong enough that I was of significant help to my dad and he depended on me to pitch in with the milking, chores and farming. When the disease kept me indoors, my mom helped outdoors. During my two years in bed, I felt guilty that I wasn't doing my share around the farm, so I tried to help with indoor chores. I became very good at ironing and cooking and to this day, I can make a perfectly lump-free gravy!

It was a cold, snowy, winter night when Lena woke up Ole.

"I tink the baby's coming, Ole, you had better go get da doctor."

There was no phone and the electricity was out, so Ole saddled his horse and rode five miles for the doctor. When they got back, Ole lit a kerosene lantern so the doctor could examine Lena.

After a couple of minutes he said, "Ole, you're about to be a father. Make yourself useful. Hold the lantern right here so I can see what I'm doing and I'll deliver the baby."

Ole held the lantern and it wasn't long before the doctor said, "Here it comes, Ole. You're the father of a baby boy!"

"Ya, sure, dat's vunderful!" said Ole.

A few seconds later, the doctor said, "Wait, Ole, hold that lantern steady. It's twins, Ole, congratulations! You're the father of twins!"

"Oh, golly!" gasped a shocked Ole.

"Don't move that lantern, Ole, I think there's another one coming!"

Ole said, "Doctor, do you tink it's da light dat's attracting dem?"

The Samuelson barnyard, circa 1915.

Chapter Two

Life on the Home Place

Often, I joke about the grass not always being greener on the other side of the fence. I tell people I decided to leave the dairy farm because I didn't want to spend the rest of my life getting up at 5 a.m. to milk cows. But ever since becoming a farm broadcaster, I have been getting up at a quarter to three in the morning to do a program for people who get up at 5 a.m. to milk cows! Just one of those small bumps in the road.

My daily routine on the farm was crawling out of bed at about five and heading out to the barn where I'd feed and milk the cows and help Dad clean the barn. After breakfast, I would change clothes and go to school. Through eighth grade, I walked the mile to and from school. During World War II, when we had Daylight Saving Time year around, I often walked to school in the dark because the sun wouldn't come up until after eight o'clock. There were mornings when it was 35 degrees below zero and we'd still walk to school. We had some winters where the snow was so heavy, Dad or Mom would take my sister and me to school in a sleigh, pulled by our draft horses, Buster and Blackie. It was the same sleigh we'd use to get milk to the milk truck when we were snowed in and the plows wouldn't get to us for three weeks. We would load cans of milk onto the sleigh and pull it to the main road.

O'Connell School, our one-room school that served about 40 kids in eight grades had a wood stove for heat. It had his and her bathrooms: two outhouses which were not heated. The school was next to a woods which provided a great setting in which to play cowboys and Indians and war and everything else. Based on all of the safety rules in effect today, it's a wonder that we survived.

When I got home from school, I didn't have to look around for something to do because it was time to throw silage down from the silo and get water to the dairy calves, clean the pens and do other work. Then it was time to milk the cows and we would generally get back into the house at about eight o'clock at night. Then it was homework and off to bed. And before electricity came on the scene, we would do homework by Aladdin lamp and kerosene lantern. (This is what my kids David and Kathryn call my "Abe Lincoln log cabin story.") We also used the lanterns while doing chores after dark and how we never burned down the barn, I'll never know. We carried that open-flame kerosene lantern up into the hay mow and it could easily have been disastrous.

In the summertime, the longer days meant more work. We'd be in the field doing everything from oat harvest to hay harvest. We didn't bale hay; it was loose hay that we would pile onto the wagon, which we'd pull

An early encounter with a horse appears to be going well as I sit on one of my Uncle Bernt's horses. But I had a love-hate relationship with our work horses.

into the hay mow with the horses.

Buster, Blackie and I never got along very well. I didn't like them they didn't like me, but they were work horses in every sense of the word. We did have a Samson tractor, which was about as clunky a machine as you would ever find — no power steering or anything like that. Years later, I found it was built in a plant in Janesville, Wisconsin that later became the General Motors plant. It wasn't very reliable and would often break down, so Buster and Blackie were our main source of power for farming on the farm. They'd pull the hay mower, the hay wagon, the plow, the cultivator. We really depended on them.

Milk was our main source of income. The milk check monthly; we raised crops to feed the cows, and believe it or not, we also raised tobacco. I get the oddest looks when I mention that we grew tobacco in Wisconsin, but there were two areas in the state, south of Madison, near Stoughton, and then up in our area, which is east of La Crosse, where tobacco grew. It was a cigar-type tobacco — binder-type, we called it — that would be used as the wrapping for the cigars. Oh, I hated raising tobacco. It was the worst crop, from a labor standpoint, that we grew. One year, I kept count and each tobacco plant was handled by hand sixteen times, beginning in April when you pulled the plant from the tobacco bed and continuing through the growing and harvesting season over the summer and fall. And then, by the next February, we would be in the tobacco stripping shed, where we stripped the leaves from the stalks, putting them in a bale. Then, the buyers would come out, they'd open the bale, pull out a leaf of tobacco, take a lighter and touch it to the leaf. If the tobacco flared, that wasn't good. If it didn't burn, that wasn't good. But if they saw a red coal slowly moving up the leaf, that was what they wanted for the outside layer of tobacco. And the leaves were fragile, too. If we got any kind of a hailstorm during the growing season, that would wipe out the tobacco, because holes in the leaves didn't work. As much of a pain as it was to grow, the reason that a lot of farmers did it, and I know in my dad's case it was the reason, was that the tobacco check would come in time to pay the real estate tax. I remember getting checks for $500, $530, for that year of hard work with the tobacco crop.

We sent our milk to a cooperative creamery that produced butter. The milk it didn't use for butter was shipped on to another factory that would do cheese. I had an uncle who had a small, rural cheese factory, and I'd love to go there and watch him turn the milk into solids, including the curds. Oh,

I was often the winning bidder for championship cheese at the Illinois State Fair.

I ate a lot of curds! They'd come flying out of the machine that would chop the cheese slabs into the curds and, yeah, I ate a lot of curds. Finally, with the use of pressure, my uncle would shape the cheddar into forty-pound blocks. My appreciation of cheese continues, and for many years, I was the winning bidder of the grand champion cheese at the Illinois State Fair.

Eggs from our 200 chickens were another source of income, but usually as barter. We'd take what we didn't use on the farm to town and trade them for groceries at Johnstone's Grocery Store.

We butchered our own hogs and we also kept one or two cattle that we butchered for beef each year. Before electricity arrived, keeping the meat from spoiling was a challenge, especially in warm weather. Mom would cook the meat and put it into Mason jars. Our water came from a cistern on the farm that was about thirty feet deep and had only about ten feet of water in it. In the summertime, we'd put the jars of meat and other perishables, like butter and cheese, into a basket on the end of a rope, and lower it into the cistern, tying it off just above the water line. That kept everything cool enough. The meat, especially the pork, contained a lot more fat than today's cuts. But, people at that time wanted bacon with fat on it, partly because of the taste and partly because they wanted to be able to render the fat so they'd get lard. We didn't have Crisco and other cooking oils at that time, so Mom would cut the fat off, render it, and we'd have lard to cook with.

We had no indoor plumbing and never did as long as I was on the farm. Dad didn't think you ought to go to the bathroom in the house. That was just his feeling, so we didn't. We had a two-holer outhouse. We described our house as having four rooms and a path.

Saturday night was bath night. We took a bath once a week, whether we needed it or not, as the old joke goes. We sat in a galvanized tub and the water would be heated in a boiler on a wood-burning stove. Kalamazoo, Michigan is forever ingrained in my mind because the stove's manufacturer, Kalamazoo Stove Company had its name on the stove in raised letters. I traced those raised letters, K-A-L-A... traced all of them over and over with my finger. I knew how to spell Kalamazoo at an early age.

We had plenty of wood to burn, so we didn't burn coal. Part of our fall work once we were done harvesting was cutting trees, trimming them and hitching up Buster and Blackie to pull the logs into the barnyard where they would be sawed and split into firewood. Just as there were threshing runs and there were silo-filling runs — where one person would own the threshing machine, one person would own the silo filler and make the runs to each farm that needed help — there was a person who owned a woodcutting rig and he would go from farm to farm. So after we would get twenty or thirty logs together, he'd bring in the rig and turn it into firewood.

That Kalamazoo stove was all the heat we had in the house and we didn't stray too far from it on cold mornings and during the really cold nights of winter. We would sleep in the front room, as we called it, on the floor near the wood heater and that way we would keep it stoked all night. When it wasn't quite as cold, but still freezing, we'd sleep in the bedrooms, but the fire would die by ten o'clock. By morning, it was ice cold again in the house, so I would take my next day's clothes to bed with me and have them under the covers so they'd be warm when I got up to get dressed.

We didn't work all the time, but there wasn't much time for leisure, nor many places to be entertained. Every month or so, we were treated to the luxury of a movie in town. Most of our socializing was centered on the church. Once we got into high school, there were basketball games each week during the winter. And on Saturday nights in the summers, the merchants of Ontario, Wisconsin — population 527 — would show black-and-white movies in the village square free of charge and keep the stores open until nine o'clock. Farmers would take their families to town, do their weekly shopping and socializing and then sit in the square and watch the movies. You had to make your own entertainment in those days. There was no Internet or iPods or 500 channels of cable television.

I missed the first half of the Great Depression and was too young to remember the rest of it, but I certainly heard the lessons my father learned from it: "Don't buy what you can't afford and don't borrow what you can't pay back." He said that over and over and over again, and he practiced what he preached. For example, when we built a new barn in 1948, he didn't buy a piece of lumber until he had the cash.

ELECTRICITY

There are certain dates in our lives that we never forget. Birthdays, anniversaries, major news events. April 11th, 1947 is a date I'll always remember because what happened on that date changed my life. Crewmen for the Rural Electric Co-Op drove their trucks onto our farm and hooked us up to electricity. Our lives changed instantly. My mother threw away the heavy flatirons that she had used to press clothes. We got a refrigerator. There was no longer a need to put perishables in a basket at the end of a rope and lower it into the cistern. I was able to stand at the foot of the stairs, flick a switch, and there was light in the upstairs bedroom. Suddenly, I didn't have to worry about rationing my radio listening for fear of the batteries going dead. We got an electric powered radio. Until that day, one of my most cherished Christmas gifts was a small wind-up phonograph, where I'd play my 78 rpm records. In 1948, I got an electric phonograph.

Outside, we no longer had to milk cows by hand; we bought a milking machine. Instead of putting the cans of milk in cold water to keep fresh, we had a milk cooler. It changed everything for us and I know that just as life on the Samuelson farm would never be the same, a similar change was happening all across rural America. The Rural Electric Co-Ops probably brought the biggest cultural change in the history of rural America. Lives were changed.

It would have happened earlier because they were ready to hook us up just before World War II broke out. But when the war broke out, copper was taken for the war effort, and so there was no wire available to wire our farm. Once the war ended, it took a couple of years before we got back to manufacturing wire and we could get the farm wired. Dad and Mom had actually purchased a refrigerator before the R.E.A. came along and hooked us up. Our most expensive purchase was the milking machine, but that was considered absolutely vital because milking was a tough job.

Some of our older cows never did learn how to take a milking machine, and it was my job to milk those cows by hand. Learning how to milk cows by hand is like learning to ride a bike. You never forget how. That's why I am a five-time Illinois State Fair cow milking champion!

Another important date happened in May of 1939. We took delivery of our first tractor. It was a shiny, red Farmall F-20. But again, the war got in the way. Rubber wasn't available, so the tractor ran on steel lugs. After the war, we took those wheels off and put rubber tires on the tractor. That was a tractor with no frills. You started it with a hand crank — there wasn't a starter on it, but compared to farming with horses, it was a great improvement. We were able to get a cultivator, a hay baler and all the other things that we didn't have before. But, again, because factories were making war materials and not farm machinery, we didn't get most of that stuff until after the war.

People often ask me which had the biggest impact: mechanization or electrification. I say it was electrification because it affected your entire life. The tractor sure made farming easier, but that was just the farming part of your life. Rural electrification was culturally and socially and agriculturally beneficial. It went far beyond just farming. It made us less isolated. Our mailbox was a mile and a half away. We had no telephone. We had no daily paper. Because of the battery rationing during the war, our radio listening time was very limited and by January of 1945, our batteries were all spent. We were virtually without communications, but there was always yelling. On April 12, 1945, when I was 11 years old, I was outside doing my chores when I heard a metallic clanging echoing across the valley. About a half mile away, I could see our neighbor banging on a washtub. When he saw that he'd gotten my attention, he yelled, "Tell your folks the President died this afternoon." That's how we found out that Franklin Delano Roosevelt had passed away.

Ole invited his boss over to the house for dinner. As Lena was

setting the table, Ole's boss was making small talk with Little Ole.

"What is your mother serving for dinner, son?"

Little Ole said, "I tink she's making buzzard."

"Buzzard?" Ole's boss said, shocked at the answer.

"Ya," explained Little Ole. "Dis morning Mama said to Papa, 'If ve are going to have dat old buzzard for supper, it might as vell be tonight.'"

1954 at WHBY in Appleton, Wisconsin with Clyde Downing. Clyde was the morning man at WHBY and before that, we'd worked together at my first station, WKLJ in Sparta, Wisconsin.

Chapter Three

From Polka to Pork Bellies

One year of college tuition was my reward for being salutatorian of my high school class, an accomplishment that isn't quite as impressive as it sounds; there were only 26 kids in my class. But it got me a one-year scholarship to the University of Wisconsin. It was a proud moment for our family because education was important. My dad had gone only as far as sixth grade. Mom went to two years at a teachers' college before they got married. Off I went to Madison to become the Samuelson family's first college graduate.

I lasted three weeks. I was homesick and I wasn't getting the education I had expected. I wanted to take broadcasting classes and learn to be an announcer, but the university wasn't teaching broadcasting; it was teaching journalism and writing. To this day, I'm not a writer. I do all of my writing with my mouth. I made up my mind that the University of Wisconsin wasn't going to get me any closer to what I wanted to do and I was ready to quit. All I had to do was talk with my parents about it, which wasn't easy in different ways. We didn't have a telephone on the farm and I couldn't just run home — it was nearly 100 miles — so I had to call relatives in Ontario and ask them to drive out to get the folks and have them go into town so I could talk to them on the phone.

It was a difficult conversation, knowing how important my being in college was to them. I told them why I wanted to quit and my dad said I should take the bus home the following weekend and we'd talk more about it. I imagine he and Mom were hoping or expecting that my homesickness would wane and I'd start liking my classes better. But before hanging up, he said, "If you can find a school that will teach you how to be a radio announcer, then we'll let you quit. But, you can't quit until you find it."

I found the American Institute of the Air in Minneapolis, Minnesota that, in six months, taught me to be a radio announcer. Among other things, I learned how to pronounce the name of every difficult classical composer in the world, which I've never had to do since. I was taught how to write commercial copy and all of the basics of being on the radio. It wasn't a cheap school, something like $1,400 for the six-month course. My parents paid for some of it, but I worked at a filling station in Minneapolis. School would go from seven o'clock in the morning until two o'clock in the afternoon, and then I would go to work at the Shell station and pump gas in the afternoons and evenings.

At the end of the six months, I received my Radio Telephone Third Class Operator Permit from the Federal Communications Commission. It was commonly referred to as a "Third Phone" and was a minimum requirement for radio station announcers at that time. It authorized me to take meter readings from the transmitter and operate a radio station.

My voice hadn't changed yet when I decided I wanted to try to become a radio announcer. That was when I was bed-ridden and spent hours each day listening to famous announcers with their rich, baritone voices on far away radio stations. There I was, with a squeaky, little boy's voice wanting to be like them. Over the years, the voice, like most every other part of your body, matures, and in radio, because it's used a lot and in different ways, the voice probably matures a little quicker. The odd thing is, I really don't hear anything different or special about my voice, but I thank God other people do. It's apparently a very recognizable voice and it's a loud one. I was blessed with a lot of volume, but I learned early in my career that I need to be careful in public places. When I was at restaurants, people would overhear my conversations and tell me, "Hey, I know that voice. I listen to you on WGN." Well, it was flattering, but it was a lesson in being careful about speaking too loudly.

The volume probably was a result of my goal to be a farm broadcaster and I knew I would have to compete with the sound of the cows in the barn being milked, the sound of the tractors and other farm noises. So, I wanted to make sure that I not only spoke loudly, but used proper diction, enunciated well, pronounced words correctly and used good grammar. Luckily, I had an English teacher in high school, Mrs. Woods, who pounded all of that into me for four years, bless her heart.

WKLJ-AM, SPARTA, WISCONSIN

Even before the six months of broadcasting school was over, I had my first job at a radio station. It was at WKLJ, a 250-watt daytime station in Sparta, only 17 miles from our farm. Every weekend I would drive home from Minneapolis and be a polka music disc jockey.

On a Friday in August of 1952, I graduated from the Institute and on the following Monday, I went to work full-time at WKLJ. I was making a whopping fifty dollars a week when I started, and before long, they turned me loose to sell commercial time. From every commercial that I sold, I got ten percent. As I recall, the price of a commercial on WKLJ was four dollars for a minute, so, forty cents for each sixty-second commercial I sold! But I was learning, and I was doing something I loved.

Since it was a daytime station and so close to home, I lived at home and would get up in the morning, milk cows, change clothes, go to town and be a radio announcer. After sign-off, I'd get home in the evening in time to milk cows again. I did that for the two years I spent at WKLJ and, in terms of my work ethic, it set the table for how I have spent many of my generally long broadcast days. There've been many days when I would get into WGN before dawn and take care of the morning drive broadcasts, hop on a plane to Washington, DC, interview newsmakers, then be back home that evening, just in time to milk the cows if I had still been on the farm!

WHBY-AM, APPLETON, WISCONSIN

The gentleman who helped me get the job at WKLJ knew my family and me from Ontario. Clyde Downing was a very good morning personality on WKLJ and about two months after he got a job at WHBY in Appleton, Wisconsin, he called me. "Hey Orion, there's an opening here for a nighttime staff announcer. Why don't you come over and apply?" I did, and moved to Appleton in 1954. That's when I said goodbye to the dairy farm.

My polka experience in Sparta came in handy. Wisconsin is polka country. They take their polka seriously and in Appleton, we would do live

This Jolly Old St. Nick may have said, "O, O, O, Merry Christmas!" WHBY, 1954 or 1955.

polka shows with some of the legends of polka: Whoopee John, who had hits like, "In Heaven There Is No Beer," and "The Laughing Song." There was Harold Loeffelmacher and the Six Fat Dutchmen, Romy Gosz and the Goslings, Cousin Fuzzy and the Cousins, Lawrence Duchow and the Red Raven Orchestra, Frankie Yankovic and the Yanks. We had all the big bands.

Although WHBY was a bigger station in a bigger market, I still wasn't making very good money. I worked evenings and when my radio shift was done, I'd go to a restaurant and help them clean up. On Friday nights, I worked at a clothing store in the formal wear department. I would like to take this opportunity to apologize for all of the ill-fitting tuxedos I was responsible for at weddings and proms.

Every Saturday night at 10:30, I had a telephone request show called "Music 'Til Midnight," where teenagers would call in and request songs, but before that there was a network show. WHBY was a Mutual Broadcasting Network affiliate and WGN was one of the network's primary stations. I ran the board for the network show, which meant I made sure the program was on the air, inserted local commercials and station breaks at the proper times. The program originated at WGN and in the final five minutes, Colonel Robert McCormick, the Chicago Tribune chairman, delivered his five minute conservative "sermon." I listened to WGN a lot because it was one of those stations that most small country deejays listened to and thought, "Hmm, it sure would be nice someday..."

WBAY-AM AND TELEVISION, GREEN BAY

WHBY was owned by the St. Norbertine Fathers who founded St. Norbert's College in De Pere, Wisconsin. They also owned WBAY radio and television in Green Bay. WBAY-TV was a pioneer in farm television. Monday through Friday, they did a one-hour live program in the studio, purely about agriculture and complete with a band. In 1956, one of its three farm broadcasters left. The sales manager at WBAY, Earl Huth, was the man who had hired me in Appleton before moving on to Green Bay. Huth remembered that I grew up on a farm, so he telephoned me.

"Orion, what do you think about working up here on the TV station and the radio station and doing agricultural programming?"

I didn't think twice. "Hey, that's right down my alley. I would be happy to."

In 1956, I moved to Green Bay where we did a two-hour farm show in the morning on the radio and then that hour at noon on television. It was a fun show. We'd drive tractors and trucks into the studio and do live commercials on the benefits of the vehicles. We had 4-H kids bring their hogs and steers into the studio and show how to feed them and how to fit them and get them ready for the show ring. We had our own band. I learned to shoot, process and edit film, skills which came in very handy in later years as I traveled the world. We would go to meetings and other farming events with a Bell and Howell 16-millimeter silent camera, shoot film and bring it back to the station where we ran it through the processor. Once it was processed, we'd splice and edit it and put it together to show on the air. When we weren't able to shoot film, we'd take a black-and-white Polaroid camera (the sensational development in 1956 was that the Polaroid would develop a picture in sixty seconds), shoot a picture of, say, the new officers of the Wisconsin Dairy Association and after it magically developed, put it in front of a TV camera — cameras that were five times bigger than the studio cameras our TV show uses in 2012. We did that for four years and I learned a tremendous amount about the programming of agriculture for television and radio.

WBAY did more television programming focused on agriculture than any station in the country. Remote television broadcasts were very rare in most cities, especially in small markets like Green Bay, but we did several, including from the Wisconsin State Fair. That was not a cheap endeavor, requiring truck loads of equipment and a lot of manpower. You just didn't find other stations doing that, but farm broadcasting was a money maker for WBAY. As a result, when we would travel to the National Association of Farm Broadcasters' meetings, anybody who was in television or thinking of doing farm television, talked to us to find out how to do it. Bob Parker, who was the farm director, gets the credit for getting it started at WBAY. He was a very persuasive man who got along very well with our general manager, Hayden Evans. Hayden understood the importance of agriculture and Bob was able to get him to commit to the programming time and money to do what we did. Advertisers climbed on board quickly and we had Ralston Purina, the fertilizer companies, the seed companies, International Harvester and John Deere, all willing to spend money to support that programming. By 1960, both Bob and the number-two-man in the farm department, Les Sturmer, had moved on to other positions and I was the farm director.

Those years in Green Bay were very important for me. I made my first trip to Chicago because the National Association of Farm Broadcasters — then known as the National Association of Farm Directors, which I joined in 1956 — had its annual meeting at the Conrad Hilton in Chicago, coinciding with the National 4-H Congress, also at the Hilton. Both were held during the International Livestock Show at Chicago's Union Stock Yards, at the 9,000 seat International Amphitheatre. The Amphitheatre was at 42nd and Halsted and was the site of many memorable events such as concerts by the Beatles and Elvis, sports events, circuses, car shows, boat shows, religious gatherings and, perhaps most memorable, several national political conventions, including the Democrats' infamous 1968 convention. As a farm broadcaster, I'd go to Chicago, take care of business at the Farm Broadcasters' Convention, cover the 4-H Congress, interviewing the 4-H kids, and then I'd run out to the International Livestock Show. After I arrived at WGN, the station provided me with a large hotel suite where I lived for an entire week and entertained

4-H kids and their parents, as well as our radio advertisers. From 1957 until 1975, I spent every Thanksgiving Day at the International Amphitheatre. It was opening day for the International Livestock Show, a very important Chicago event.

It was during this time that I met another Wisconsin farm boy, Larry Caine. Larry Caine worked for the Union Stock Yards and primarily managed all the events at the amphitheater as well as doing livestock reports every day at noon on WGN radio from our broadcast booth at the Stockyards. Larry impressed me with his sense of humor and his ability to manage livestock and people, including the Beatles. He introduced me to Bob Hope, Dizzy Dean, Roy Rogers and Dale Evans, when they appeared at the International Livestock Show. I also learned he was one of several people who recommended me to Ward Quaal when WGN was looking for a new Farm Director. Larry and I were lifelong friends and worked on many projects together. The stock yards were torn down in the summer of 1971 and it was just a few years later that the International Livestock Show ended in Chicago and moved to Louisville, Kentucky where it still happens every November.

Also while working at WBAY, I took my first trip to Washington, DC, where I got to meet President Dwight D. Eisenhower and Secretary of Agriculture Ezra Taft Benson, who later became president of the Mormon Church. The Green Bay years were a very exciting time in my young career.

Ole and Lena had lived a long life together, but finally, Ole's health failed to the point that he had little time remaining.

Lying on his death bed in an upstairs bedroom, Ole smelled the delightful aroma of lefse being baked by Lena in the downstairs kitchen. He rolled out of bed, and being too weak to walk, climbed backwards down the stairs, crawled to the kitchen table and reached up to enjoy one last piece of lefse.

Lena saw him, walked over, slapped his hand away and said, "Ole! You can't eat this lefse. It's for the people coming to your funeral!"

Live on WGN Radio from the International Amphitheater and the Chicago Market Jamboree in 1962.

WGN Radio

Chapter Four

WGN

"Orion, did you hear what happened at WGN?" It was 1960 and John F. Kennedy was running for president. Bob Parker, who had been my boss at WBAY before moving to Chicago in 1959 to join an advertising agency, was on the phone asking me if I'd heard that WGN Radio Farm Director Norm Kraft had resigned. Norm, during the *Country Fair* noon show on WGN, resigned on the air, announcing he was going to work for the Kennedy campaign. Norm's hope was that JFK would win and he would name Norm as his Secretary of Agriculture. JFK won, but Norm did not become secretary.

"He quit and walked out with no notice," said Bob. "WGN is in a hurry to get somebody. I just talked to Ward Quaal and I put your name in the ring."

"Oh, Bob, I'm not ready for WGN," I said.

"I think you are and I'm not the only one. Layne Beaty at the Department of Agriculture in Washington has also called Ward to get your name on the list."

Meanwhile, the following Monday, the general manager of WBAY called me into his office.

"Orion," he said, "We've been thinking about this and if you're worth a thousand dollars a month to us a year from now, you're worth it right now. So, I'm raising your salary now."

He didn't know that I knew my name was in the mix for the WGN job, and I didn't let on that I did.

Actually, I was beginning to wonder if I really was in the mix because days went by and nothing, not a word from WGN. I wasn't really interested in moving, but I thought to myself, "Maybe I should apply just to see where I stand in the farm broadcasting industry." I'd only been at it four years at WBAY and looked up to the big names, the stars of agricultural broadcasting: Herb Plambeck at WHO, John McDonald at WSM, Bob Miller at WLW, Maynard Speece at WCCO and others.

Finally, ten days after I was told my name was among those being considered, Bruce Dennis, the program director at WGN called.

"Orion, we'd like to talk to you. Will you come down to Chicago?"

I agreed, telling my wife, Nancy, "This isn't going to happen. We've just built this new house (a two-story home for $24,000), we're very comfortable here, but you know, at least it's a free trip to Chicago." I also said, "I just know that people in that market are cut-throat. If you make a mistake, there are ten other people ready to take your place."

So, at WGN's expense, I flew North Central Airlines to Chicago, looking forward to seeing the station I had admired and listened to for years. Everyone I met at WGN — from Ward Quaal to Eddie Hubbard, the morning man, to Wally Phillips, who was doing evenings then, to Jack Brickhouse — everyone couldn't have been nicer and I was very impressed. At the end of the day, Ward said, "Orion, we'd like to hire you." Suddenly, a day trip to Chicago had become a potential life-changing event. I told him, "Well, I've got to go back to Green Bay and talk to my wife." They gave me a deadline by which I had to let them know and I went home.

I was shocked when they offered me the job. I really thought long and hard about it and it took me ten days, and another trip to Chicago, to make the decision. I really enjoyed Green Bay. I was a big fish in a little pond at WBAY, and was making a name for myself and building a nice life. The WGN deadline had arrived and I called Bruce Dennis.

"Bruce, I just don't know."

"Tell you what," he said. "You and your wife come to Chicago as our guests, take a look around the city, look at housing and all that sort of thing." And we did. And when we got back to Green Bay, I took another day before deciding. "This is dumb," I told her. "Put WGN call letters on my resume and I can go to any station in the country. So yes, let's do it for five years, come back to Wisconsin, buy a radio station and I'll spend the rest of my life running it." I look back at that now, and shudder to think how close I came to saying no.

When it came time to negotiate my salary, Ward Quaal said, "Orion, what do you think you need to make the move?"

I said, "I have no idea."

"Well, give me a number and if you're high, I'll tell you, and if you're low, I'll tell you." He was just that kind of person, the kind you'd believe in and trust.

I said, "How about $14,000 a year?" That was $4,000 more than I had been making in Green Bay.

"Well, I think we can do that, Orion." He told me that in six months, they'd take a look at my work and probably adjust it because he expected that I'd do very well.

"What about a contract? Do I need a contract?" I asked.

"Orion," he said, "I'm from Ishpeming, Michigan and you're from the Kickapoo Valley. A handshake works in Ishpeming and I'll bet a handshake works in the Kickapoo, too." We shook hands and 24 years later, the attorneys in the Tribune Tower were shocked to discover that I didn't have a contract. They changed that in a hurry!

When I moved from Green Bay to Chicago, it was a major adjustment from a lifestyle standpoint. For one thing, I rarely had to use my brakes in Green Bay. On Chicago's freeways and busy streets, I was wearing them out! We lived in Evanston for about six months, then bought a home in Glenview and a few more over the years in other areas of suburban Chicagoland. To be on the air at five o'clock, I was getting up at 3:30 a.m. After Nancy and I divorced and my second wife Judith had died of Lou Gehrig's disease, Gloria and I were married in 2001 and we moved to Del Webb's Sun City community in Huntley, 52 miles from the Tribune Tower studios. I had to set my alarm even earlier, at a quarter to three in the morning. Early hours aside, it didn't take long to find out that this was an exciting career and an exciting job.

September 26th, 1960 is the first time I said, "This is WGN Radio, clear channel radio serving the nation from Chicago." My first show was *Milking Time* from five to five-thirty each morning, Monday through Friday. At noon, we did an hour called *Country Fair*, which later became the *Noon Show*. I did market reports during other hours of the day and I quickly learned that my concern of losing my job when I made a mistake was for naught. The station was in the Tribune Tower; both radio and television were there. The studios were on three different floors and my office was on the first floor. I'd have to go up to the third floor to do the reports and the first two or three weeks, I missed three market reports because I was waiting for an elevator.

On April 30, 1960, WLS-AM fired nearly everybody on its staff. It had been the Prairie Farmer Station and carried the very popular *WLS Barn Dance*, but it signed on the next morning as *Wonderful Rock and Roll WLS*, and they became the first rock 'n' roll station in Chicago. Gone were all of its musicians and the people who were doing agriculture reporting. It was no longer the Prairie Farmer station. That infuriated farmers across the country, but switching to rock 'n roll worked out well for WLS, of course. It also created an opening and Ward Quaal made a decision to fill the void, increasing the amount of airtime devoted to agriculture. After I got to WGN in September of that year, I told Ward, "You know, the great talents who were fired, they're still out of work. We ought to bring the *Barn Dance* over to WGN." He agreed and from 1961 until 1969, every Saturday night I was at the WGN studios to announce the *Barn Dance*. We'd start rehearsal at three

With Dolph Hewitt and the Sage Riders: (L-R) Lino Frigo, Tiny Murphy, Toby Nix, Jimmy Hutchinson and Dolph.

WGN Radio

Part of the WGN Barn Dance cast - Donald "Red" Blanchard (funniest story-teller I've ever known) and Bob Atcher, country singer who appeared in several Western movies. I worked with them every Saturday night on radio & TV in the '60s.

o'clock. We'd have two hundred people in Studio Three and we would do a live television show, 6:30 to 7 p.m. Then, we'd dismiss that audience, bring in a new audience and at 8 p.m., do a 90-minute radio version of the *WGN Barn Dance*.

The *Barn Dance* had a big audience and attracted some of the best singers, bands and entertainers in the country. I worked with Donald "Red" Blanchard, Arkie the Arkansas Woodchopper, Lulu Belle and Scotty, Dolph

Hewitt, Bob and Bobbie Thomas, the Johnson Sisters, Bob Atcher and many more. Bigger names, like Eddy Arnold, Minnie Pearl and Johnny Cash, came to Chicago to do the show. The night Johnny performed, we had to lead him on the stage and lead him off. He didn't quite know where he was, but when he was on microphone, he was perfect.

On the WGN radio noon hour shows, we would do two or three market reports from the Chicago Board of Trade and the Chicago Mercantile Exchange. The Chicago Union Stockyards was a vibrant livestock market with thousands of cattle and hogs coming in by truck and by rail everyday, so we had a broadcast booth out there. Two employees of the stockyards, Larry Caine and Bob Kuhn, would help us out. One or the other would go into the booth and go live on the air each day at 12:15 during our *Country Fair* show. They would tell us who had been selling livestock and who had the top-priced cattle that day. That was juicy news for rural listeners who might hear their neighbor's name mentioned and say, "Well, so-and-so did pretty well; he'll have to buy dinner when he gets back!"

Country Fair and *The Noon Show* had a live orchestra and a vocalist every day. Our bands included Eddie Muhm and the Country Fair Boys, a five piece band, and on the one day a week that we'd have to give them the day off, we had the Bob Trendler Bozo Circus Band — 16 pieces — doing the farm show! Elaine Rogers was one of our wonderful "girl" singers. Elaine was best known for singing a jingle for Budweiser, the one that went, "Where there's life, there's Bud." Elaine was well paid from that because Budweiser used it over and over again for many years. Elaine and the band would travel with me and, oh, did we travel! We had a tremendous support staff of engineers in those days. We did 40 to 50 remote broadcasts a year. We would do eight days at the Illinois State Fair; we would do five days at the Wisconsin State Fair; we'd go to the Indiana State Fair. We went to all of the big outdoor farm shows, too; the Farm Progress shows, the Farm Power shows, Husker Harvest Days, the Sun Belt Expo and more. Those remotes were big productions. We would take two, sometimes three engineers along and a truckload of equipment. Now, in 2012, I can do my show on a cellphone driving down a country road anywhere in the Midwest.

The remote broadcasts were exciting and I probably met more of our WGN listeners personally than anyone else on the staff because we were out there with them all the time in all sorts of places. Carole March, who was one of the singers with Lino Frigo and the Musical Wheels remembers that we did a *Noon Show* at a veterinary convention in St. Charles, Illinois. The veterinarians were performing surgery on a sow on the floor in front of our stage while we were on the air. She remembers being on stage while the sow was getting a hysterectomy. "I was singing and looking out at this 300-pound sow and we all looked at each other and started laughing," she said. Carole was just one of the countless talented performers I worked with over the years. Others included Lola Dee, Cathy Johnson and Anita Roman. The bands included Dolph Hewitt and the Sage Riders, Captain Stubby and the Buccaneers, Lino Frigo and the Musical Wheels, The Eddie Vodicka Orchestra and many others; just a lot of very talented people.

The early '60s were a big time for radio. Timing, as the saying goes, is everything and I was in the right place at the right time. It was great fun and I was known throughout the country because of what we did on WGN.

The WGN call letters became so powerful that when I started going to Washington, DC to chase down interviews and stories about agriculture, just saying I was from WGN meant instant access. When I walked into a senator's or representative's office, everyone knew right away who I was and what I was doing. Ultimately, it led to several White House interviews and to dinner at the White House. What an eye-opener that was for me. It was very thrilling to be among the people making decisions for the whole country and to this day, I get excited every time I fly into DC. Working at WGN led to things which, as that country boy on a farm in Wisconsin, I could never have dreamed.

Because this job is so much fun, I haven't worked a day since I got into broadcasting. As my childen, David and Kathryn, were growing up, I told them, "Find something you like to do so much you'll do it for nothing. Then, do it so well you will get paid for it." That is what I've been lucky enough to do.

But, leave it to your parents to bring you back to Earth. My dad and mother were never comfortable coming to the city, particularly my dad. In the 35 years I worked at WGN while Dad was alive, they came down to the city only four times. During one of their first visits, I had Dad sit in the studio with me during the *Country Fair* show. I used all the big words I'd ever learned and, off the air during commercials and news breaks, I talked all the radio talk I could with the producer and engineer. I was trying to impress him. Dad was a quiet man, he didn't talk a great deal. And he didn't have much of a sense of humor, although he unintentionally said some funny things. He was just a hard working Norwegian dairy farmer. When the show ended, he didn't say anything. No surprise, I was used to it. As we left the studio and walked back to my office, he didn't say anything. Again, no surprise. We got into the office and still, he didn't say anything. Finally, I looked at him and asked, "So, Dad, what do you think?" Serious as could be, he said, "You know, it must be nice to be able to look at all that hard work and then just talk about it." I will never get a better job description. That's what I do.

Because of how clearly unimpressed he appeared to be with my line of work, I never dared tell him what I paid for my first home after moving to Chicago. When he and Mom sold the 200-acre dairy farm in 1964, the land on which he was born and spent all of his life, they walked away with $28,000. That was for the machinery, livestock, land... everything. Dad would never have understood the amount, nearly ten times as much, that I paid for my first home in Glenview.

THE WGN FAMILY

I've had the privilege of working with the smartest, funniest, most talented and most compassionate people in all of broadcasting. Of course, over the 52-plus years I've worked at WGN, there have been hundreds of them. To list them all is impossible. To say as much as I would like to about each who has had a significant impact on my life and career is also impossible. From Eddie Hubbard, the morning man when I arrived in 1960 to Holland Engle, the newscaster on the Mutual Broadcasting System who had been filling in on the agriculture reports after Norm Kraft quit and was my mentor for the first two or three weeks, to many, many others.

I will mention some of the "old-timers," if they'll pardon me for calling them that, like Carol Hankner, who started as a sales assistant in 1964 and for the past eight years has been the executive assistant to the general manager and program director. Marlene Wells, our promotions director for most of the last 37 years, started as a secretary in the art department in 1967. Jimmy Lucas has been setting up and engineering remote broadcasts since he came to WGN in 1970. Jim Carollo started at about the same time, semi-retired in 2010 as chief engineer but came back to oversee the station's 2012 move to the seventh floor. Two others still on the job, engineer Jim Holland and the incomparable Milt Rosenberg, started in 1973. But, regardless of how long they worked there, the WGN family treated me with kindness and respect and provided me with all the support and tools I've needed to do my job.

However, you may have heard me mention on the air a few times during the period in 2010 when my 50th anniversary was being celebrated that there were only two people in all my years that I did not enjoy working with and that I would name them in "the book." Well, this is the book and those two people are... going to remain nameless. I've decided that there's nothing for me to gain by telling you their names and, besides, they know who they are. (And again, no, they aren't Kathy and Judy, as John Williams jokingly suggested during an interview he did for my 50th WGN anniversary.)

Among the many co-workers with whom I have enjoyed working, there are several I feel compelled to talk about at length.

WARD QUAAL

Ward was overly kind through the years saying that I was one of the best hires that he ever made in the industry. In September 1960, he told me, "I've hired you to give me the best agricultural radio programs in the country." He said, "If that means you're in Dallas tomorrow and Washington, DC the next day, and Denver the next day, you go. You don't come and ask me. I'm hiring you to make those decisions." He went on, "The other thing I want you to know is that WGN does everything first class — we are *the* radio station in the United States and we do everything first class." That included flying, and so I was enjoying a very exciting life for a young man fairly fresh off the farm, doing a lot of traveling to interesting events and places. And wherever I went, the WGN call letters were recognized with respect and excitement, especially in the Midwest, where the big WGN signal covers seven states in the daytime: Illinois, Iowa, Wisconsin, Michigan, Indiana and a little bit of Missouri and Minnesota. At night, with the sky wave signal, we can get into most states and a good part of Canada. We were an important information and entertainment

WGN Radio

Ward Quaal

resource because people in such a huge part of the country could hear us. We were one of 25 clear channel radio stations in the country, which meant that there was no other station on 720, we were the only one there. In the '60s, Congress started to break down the clear channels, deciding that we needed more radio stations and those 25 frequencies needed to be shared. Ward, who was a strong advocate of the clear channel system, told Congress that WGN needed to remain a clear channel station because we were the only information service for people in many states when their local stations were off the air and we boomed in as a clear channel. But that all changed.

As a matter of fact, the first one to come on our frequency was Fidel Castro. During the missile crisis in 1962, when we were blockading the Russian ships from bringing missiles into Cuba — a very tense time — President Kennedy asked four radio stations in the U.S. to carry him addressing the Cuban people in Spanish for five minutes every hour on the hour from midnight to five a.m. for four nights, and WGN was one of the stations he selected because we had a strong signal into Cuba. Kennedy's messages were aimed at the Cuban people. "It's not you we're against, it's your leadership, and we can't allow these missiles to come into your country." Well, Castro got us. Within six months, Castro had built a 200,000-watt station, which is four times more powerful than WGN or any AM station in the U.S. The Cuban station was broadcasting on all four frequencies of the stations that had carried JFK. At night, if you lived on the south side of Chicago, you couldn't get WGN, you got Cuba. It took about 11 years working through the Swiss embassy to get that changed. He moved ten cycles or so off our frequency, which helped our signal. But credit Castro with being the first person to start breaking down the clear channel radio system. Today we share 720 with other stations, including ones in Las Vegas and Connecticut. By the way, there's great irony in Ward's defense of the clear channel radio system. After Congress de-regulated the radio industry in 1996, Ward was very critical of the changes which allowed huge companies to buy thousands of radio stations, most notably one that named itself Clear Channel.

Ward started as a staff announcer at WGN, fresh out of college at the University of Michigan, and as he worked his way up the ladder, was a driving force behind the construction of the new WGN broadcast center on Bradley Place where WGN-TV is now. It was in the process of being built when I arrived. I worked at the Tower for about six months before all of us, radio and TV, moved to the very modern facility at Bradley Place. That's where

Bozo's Circus and the *Phil Donahue Show* started and where we did the *WGN Barn Dance* after WLS dropped it. After I'd been on WGN Radio for a couple of years, Ward said, "We should be doing agriculture on Channel Nine." So, we started doing specials — an hour special, a two-hour special — which took a lot of production work and travel with a camera crew to various parts of the country. In 1963, we started a daily show called *Top O' the Morning* that ran from 6:30 to 7:00 and that program continued until we moved the radio station back downtown to the Tower in 1986. From 1964 to 1986, we did a half-hour show five days a week and had a good audience. We did the show in Studio 3, which had a built-in pipe organ and an organist. Harold Turner provided wakeup music and conversation. He did for me what Ed McMahon did for Johnny Carson.

Ward went on to write a book about broadcasting that became a textbook at colleges across the country. He was an icon in the industry. When Ward took over WGN in 1956, it was a time of great change in the radio industry. Soap operas and other staples moved to television. The changes Ward made to WGN Radio were copied by stations in market after market across the country. He put traffic copters in the air to get better traffic reports. He was an outstanding, visionary broadcaster who ran WGN Radio and Television until he retired in 1975. Ward turned the call letters, WGN, into some of the most loved and respected call letters in America. He hired people who became broadcasting legends, like Wally Phillips, Bob "Bozo the Clown" Bell, Ray Rayner, Roy Leonard and many others.

Ward had as much influence on my career as anyone and he was a great friend. He passed away in 2010 at the age of 91.

WALLY PHILLIPS

When I first heard Wally Phillips, he was doing a late-night show on WLW-AM in Cincinnati, another 50,000 watt clear channel station that reached into Wisconsin where I was working. Wally was a pioneer, introducing program elements that have since been used by every radio personality and disc jockey in the world, things like sound bites and the use of listener phone calls and prank phone calls. Wally had the fastest mind of anyone I've ever known. It absolutely amazed me — listening to him take phone calls, his quick comebacks were incredible. I don't care what the caller said, he had an answer that was usually hilarious. He was very good at it and got into trouble

October, 1978 -- Orion and Jack Young (center) showed Wally Phillips how corn is harvested in Illinois. Phillips controlled 120 acres of corn scattered throughout the state, which produced nearly $30,000 for WGN's "Neediest Children's Christmas Fund."

WGN Radio

a few times. The FCC probably passed a rule or two because of Wally. He'd call people on the air and not let them know they were on the radio. It did get a little interesting at times. One of his bits was calling random numbers with an oddball or outrageous request to see what kind of response he could evoke. When the song "The Girl From Ipanema" was popular in the '60s, he dialed numbers in Brazil, asking each person who answered, "Is there a woman there who is tall and tan and young and lovely?" Or, when he called a store a few days after New Year's, insisting that they allow him to return his natural Christmas tree.

Just sitting in the studio and watching Wally and his system work was an education. In those days all of the sound bites or voice cuts he played were from a disc. He'd pull the "drops" as they're called in the business, from the dozens of discs he had in the studio, and his record turner, Fred Keller, did a tremendous job. The audience response was phenomenal. At his peak, Wally's audience was estimated at 1.5 million people. There were few, if any, local radio personalities more successful than Wally. He had more listeners in the eight o'clock hour than any radio station in the country, and that included those in New York City. There was a time that WGN had more listeners in Milwaukee than Milwaukee radio stations had.

Wally was totally focused on his work and he had to be, to be as good as he was, but it was to the point of him developing a persona of being aloof. When the show was over, he often wanted to be alone, which I can understand because anybody who put in as intense of a four hours of radio as he did would want some time to unwind. It was sort of a rule when Wally was sitting in the cafeteria after his show, you left him alone because he was still coming down. He preferred to work within the four walls of the studio where people could hear his magic, but not see it. Wally probably would not have enjoyed working in the Showcase Studio on Michigan Avenue.

Wally and I had an interesting relationship because — and later in life, we had different memories of this — for probably my first fifteen years at WGN, Wally tolerated me and that was about it. He was not unfriendly, but he just didn't understand why a station in downtown Chicago would be doing an agricultural program. He commented a couple of times that rural stations should be doing that programming, not a mighty, big-city station like WGN. I worked pretty hard to get Wally to embrace it, but without much success until the Illinois Farm Bureau decided that it wanted to contribute to the Neediest Kids' Fund, the charity that Wally started. The Bureau said it would contribute something like $10,000 by having its members each sell an acre of corn and give the proceeds to the fund. Part of the deal was for Wally to go with me to a farm at harvest time. The plan was for him to visit with a farmer, get some photographs taken and so on. He somewhat reluctantly agreed to do it because it was generating money for his pet fund, and it turned into a bigger money raiser than expected. Illinois corn farmers kicked in nearly three times the amount promised. After that he warmed up a little bit toward what I did. Later in life, I kidded Wally a lot that he tried to get me off WGN and on some other station and he'd say, "No, no, that's not true, I never did." In 1994 and 1998, Wally sent me two of the most beautifully written letters I've every received. They are treasures and are included in a "Letters" section toward the back of this book.

Wally Phillips was the undisputed king of morning radio. Wally was WGN. Part of his success, and the success of everyone on the station since then, was the involvement of the listeners. The audience developed a sense of ownership to the extent that they felt they should be involved in any management decisions. It's a culture unlike any other.

Wally battled Alzheimers for the last five years of his life. He passed away in 2008 at the age of 82.

BOB "UNCLE BOBBY" COLLINS

When Bob Collins arrived in Chicago to do nights at WGN, I honestly thought it was the end of the station as we had known it. He and his on-air sound were so foreign to what I considered the WGN culture to be. There was a certain sound and style that had been developed by superb talents like Wally, Roy Leonard, Eddie Hubbard, Jack Brickhouse, Jack Taylor and Virginia Gale who were on the air throughout the day, and Franklyn MacCormack was on all night with his *Meisterbrau Showcase.* They all had the "WGN sound." Then along comes this raucous character with an off-the-wall sense of humor and an annoying, nasally cackle for a laugh. He wasn't smooth. He didn't have a good radio voice. I thought, "My golly, what has happened to this radio station?" I didn't like him on the air for a good six months and I didn't get a chance to know him personally because he worked nights and I was generally on my way home before he arrived at the station.

After Bobby took over for Wally on the morning show, I had a chance to get to know him better and his style grew on me. Off the air, he talked about his childhood and his background. There were some tragedies in his childhood that we talked about. Then he started asking questions about agriculture, like when the sweet corn was going to be ready. On the air, he would engage me in conversation about many different areas of agriculture that, as a city person, he didn't understand. It gave me an opportunity to be a part of his program that was more than just the allocated time for the reports.

Bobby didn't much enjoy being in front of an audience. I learned that quickly when I took him with me to the annual Corn and Brat Fry in Walworth, Wisconsin. We were doing a live *Noon Show*, just as we'd done so many times over the years. A big crowd had gathered around the stage where our *Noon Show* band was playing in between my market reports and farm news. Bob was just kind of standing in the audience by himself, so I called him up on stage, telling the audience, "I want you to meet the new WGN morning guy." He came up and said hello and I could see he was not at ease; he fidgeted nervously. When the show was over, he pulled me off to the side and said, "Don't you ever do that to me again!"

Bob Collins, when he was WGN's evening personality, circa 1980.

"Do what?"

He said, "Don't you ever get me up on stage in front of a crowd. I'm very uncomfortable with that."

On stage, and even at parties, he just wasn't comfortable when he came to WGN. He grew into it because the job required that he be on stage for sponsored events, but he clearly didn't enjoy it at first. Yet, Uncle Bobby's fans loved to see him at events, and he did grow to enjoy them, especially if they were events to benefit a charity. One that we particularly enjoyed was Pumpkin Fest, staged annually on the suburban farm owned and operated by Jim and Esther Goebbert. Every October, we would do a day-long broadcast from the farm and several thousand listeners would bring canned goods for the Chicago Food Depository.

Another time of discomfort for Uncle Bobby was when I took him to a cow-milking contest at Golf Mill Shopping Center. "Bob, they want you to sit down and milk this cow for two minutes."

He took a look at the cow and said, "That thing is ten times bigger than I am. I am not going to get close to that animal!" So, I had to sit in and do the milking for him.

The best compliment I could give Bob was he could be a total clown one minute and the next minute talk gun control, abortion or other serious topics and not lose any credibility. He could make that switch from buffoonery to serious and never lose a beat. That's a rare talent.

Speaking of rare talent... when I was a child, I won a singing contest in Wisconsin doing Yogi Yorgesson's Scandinavian accent novelty songs, "I

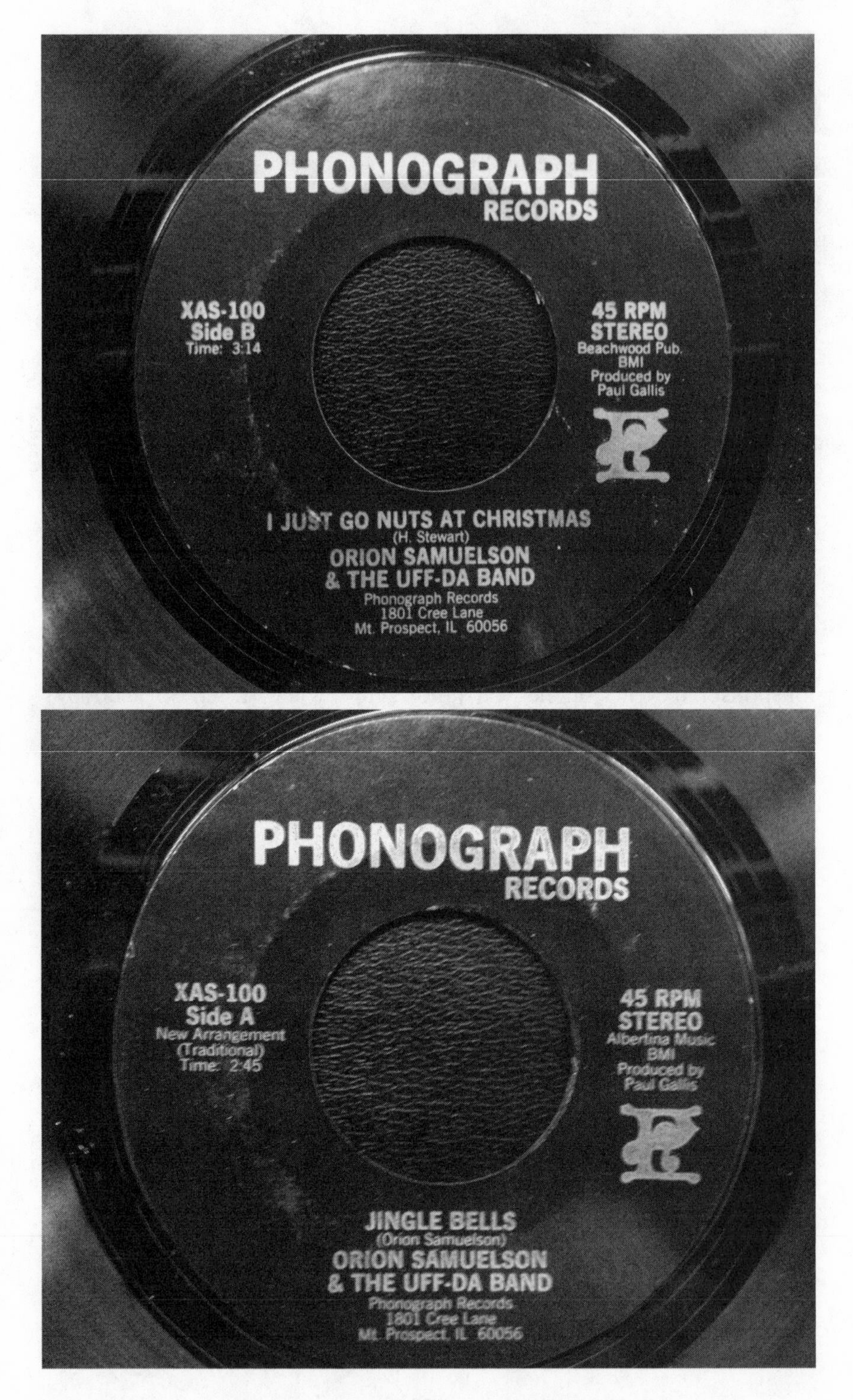
PHONOGRAPH
RECORDS
XAS-100
Side B
Time: 3:14
45 RPM
STEREO
Beachwood Pub.
BMI
Produced by
Paul Gallis
I JUST GO NUTS AT CHRISTMAS
(H. Stewart)
ORION SAMUELSON
& THE UFF-DA BAND
Phonograph Records
1801 Cree Lane
Mt. Prospect, IL 60056
PHONOGRAPH
RECORDS
XAS-100
Side A
New Arrangement
(Traditional)
Time: 2:45
45 RPM
STEREO
Albertina Music
BMI
Produced by
Paul Gallis
JINGLE BELLS
(Orion Samuelson)
ORION SAMUELSON
& THE UFF-DA BAND
Phonograph Records
1801 Cree Lane
Mt. Prospect, IL 60056

Yust Go Nuts at Christmas" and, "Yingle Bells." I won 10 shiny silver dollars, which was a big deal to young boy. Years later, when Bob was doing mornings at WGN, I was on a *Noon Show* remote from a Farm Bureau convention. We had a band with us, Lino Frigo and the Musical Wheels, and when somebody at the convention mentioned that I had won that singing contest, the band and others goaded me into doing a reprise. Of course, the engineers back in Chicago had the recorders going and the next thing I knew, Uncle Bobby was playing the songs on WGN. We started getting calls from listeners asking where they could get a copy of the songs. Bobby and I talked it over and decided to make a record. We hired a band, went into a studio and recorded "I Yust Go Nuts at Christmas" on one side of a 45 rpm record and "Yingle Bells" on the flip side. Paul Gallis, a record promoter par excellence (back in the days when we had records and record promoters) told us he would get the records duplicated, labeled and would market them for us. We made a deal to give the Salvation Army one dollar for each record sold, and we sold about 12,000 of them.

I never had much of a chance to play sports growing up, thanks to my bone disease. And, golf? There wasn't a golf course anywhere near our farm in Wisconsin. Uncle Bobby and I decided we would learn how to play golf. We were scheduled to travel to Hawaii together and we were going with two friends who were golfers. They said they were going to be playing on the Royal Ka'anapaali Golf Course on Maui, which sounded impressive and challenging, so we decided we'd better take lessons before our trip. We signed up for a crash course with Emil Esposito, who's a well-known golf instructor in the Chicago area and across the country.

The course in Maui was gorgeous, as are most of the golf courses in Hawaii are. It was laid out along the beach with mountain peaks in the distance and lush, green foliage all around us. It was a gorgeous place to play golf. Unfortunately, that's not exactly what Bobby and I did. On our eighteen holes, I had a 156 and Bob had a 154. The golf pro gently suggested that the next time, we ought to try another of Maui's fine courses because we didn't play quite fast enough, or well enough, for his course. One problem on the Ka'anapaali course was that no matter where the ball landed, it seemed to roll toward the ocean.

Despite that rocky introduction to the game, I decided it would be nice if I could learn how to play golf, so when we got back to Chicago, I signed up for more lessons. My son, David, was a good golfer in high school, in fact, he played in the state tournament and I thought it would be a nice father-son thing to do. I went back to Emil, whose motto is, "Making Golf Fun, Not Work." My problem — well, my main problem — was I couldn't hit the ball off the tee. After about the seventh lesson, I teed up, took a mighty swing and hit the ball about 25 yards. Emil glared at me and said, "Why can't you hit that ball, Samuelson!?" He added some colorful words to that question. I knew then that my golf career was over. Now I live vicariously through David and my 16-year-old grandson Matthew, both of whom are very good golfers. In fact, pardon my bragging, but Matthew scored his first hole-in-one in August of 2012. Gloria, who plays every week, is also a good golfer, but it just isn't my game.

Bobby didn't play much golf after that Hawaiian trip, either. His passion was airplanes. On the day Uncle Bobby died in a mid-air plane collision, he was my closest friend at WGN. Since we both had homes in Arizona he always told me, "When we retire, we'll sit on the front stoop and watch the desert go by." Sadly, we never got that chance.

SPIKE O'DELL

Spike found himself in one of the most difficult situations any broadcaster had ever faced — he had to replace Bobby when we lost him in the

Spike and I on a remote broadcast shortly before he retired.

plane crash. It happened during Spike's shift and he was the voice of the WGN family as we reported on it, talked about it and grieved on the air. I couldn't do it. I wasn't able to discuss it without breaking down. I'm still amazed that Spike handled that so well. There was a period of time where he knew, most everyone in the station knew, that Bob was dead, but he couldn't talk about it on the air until it was made official by the coroner. And, then to turn around and have to do Bob's show the next morning was the ultimate challenge for a broadcaster. Spike was a great guy to work with, too. Spike was a down-home boy, more so than Bob, really. He just had fun every morning when he went on the air. When people would ask me what Bob was like off the air, I'd tell them he wasn't very comfortable in a crowd and Wally was kind of in his own world, but Spike? He was the same guy. There was no difference between his life on the air and his life off the air. But he could be so goofy! One morning a listener who had dropped his cell phone in a swimming pool called and asked Spike what he should do with it. Spike told him he should put it in the clothes dryer. The caller didn't think he should do that, but Spike eventually convinced him to. We could hear the clanking noise the phone was making as it went round and round in the dryer. Spike told him to just leave it in there, it would dry out.

I asked Spike to write a few memories of our time on the air together at WGN:

I was 34 years old when I moved from Moline, Illinois to work in the big city at WGN, and I wasn't quite sure if it would ever work. Orion became a role model and mentor for me, someone to talk to if I needed to. (And, man, did we ever have some conversations!)

What I will always remember about Orion is just how much he loved to be "ON." On the radio, on television, on stage, on the spot. Most of us who were lucky enough to have worked in this whacky business seemed to always want to be around the mic, or the stage, or the spotlight. But most of us were never really good at it. The Big O was an exception. He rules the stage! His voice commands your attention! He seems completely at ease and in control of every situation, which is not an easy thing to do. (I never got that chromosome at birth, I guess. People looking back at me always scared me and I turned down every opportunity to emcee an event because of outright fear!)

Whenever I had the chance to be around Orion, it always amazed me how well he worked the crowd. Wherever we were on the "Manure Tour," be

WGN Radio

Longtime morning man Spike O'Dell in Studio A near the end of his run at WGN, 2008.

it State Fair or simply a cocktail party, the Big O was a ROCK STAR! People were drawn to him. And the coolest thing of all is that in everything he does, he does it as a gentleman and a professional.

Right before I decided to walk away from my life in radio, Orion admitted to me that when I started he didn't really care for me much and was sure I wouldn't make it at WGN! He said he was glad he was wrong about that and pointed out that he had said the same thing about a young, brash guy who came to work there years earlier named Bob Collins.

I know I can speak for Uncle Bobby when I say Orion Samuelson was, and is, a great friend.

ROY LEONARD

Roy Leonard, in my book — and this is my book — is the best celebrity interviewer of all time. Roy hosted mid-days on WGN for 31 years and did exhaustive research. He prepared — he actually read guests' books — and it showed. I'll always appreciate Roy for the gracious way he introduced me to the big stars who were in the studio when I went in to deliver my reports. He didn't have to do that, but he always went out of his way to introduce me to famous people like Gregory Peck or Neil Diamond. I'll never forget the moment when I shook Sophia Loren's hand!

Roy was kind enough to share a few thoughts about our many years together at WGN:

I remember Orion as the only guy who worked harder and longer than I did. If I was to interview an author, I'd often be up 'til 2 a.m. reading the book. But with Orion, he'd work 'til the markets closed, hop on his airplane to get to a speaking engagement, get home in time to get a couple of hours sleep and be in the studio at 5 a.m. to start another long day. He was always well prepared and knew his subject to the core.

Orion was also a generous person, willing to share his knowledge with anyone. He appealed to a broad audience including my wife, Sheila, who often made out the week's shopping list (for a family of eight) after listening to his market reports.

His holiday parties were legendary, lasting two nights so his friends in agriculture could be feted one night and the broadcasting industry the next evening. Orion and his wife, Gloria, are perfect hosts and although they have legions of friends, they remember something about everyone, so it's a lot more than just being polite.

In an industry that has turned so cold and impersonal, Orion Samuelson is a beacon of light that shines with warmth and sincerity.

STEVE AND JOHNNIE

When I arrived at WGN in 1960, our overnight program host was Franklyn MacCormack, who had more ladies in love with him and his voice than anyone I've ever known. He also sold a lot of Meister Brau beer and enjoyed some Early Times during his six-hour nightly stint, the proof of which could occasionally be found in the studio wastebasket.

But for 25 years, my radio companions on the early morning drive to WGN were Steve King and Johnnie Putman, the husband and wife team that had a huge following across the nation with our strong nighttime signal. While I didn't always agree with their politics, I did enjoy their interesting interviews with well-known musicians and their knowledge of music, enhanced by the fact that Steve was a talented guitarist.

Johnnie started her WGN career as a traffic reporter on Uncle Bobby's show and when Steve joined the WGN staff, they moved into the all-night chairs. At my WGN 50th anniversary party, they surprised me with a song written by Steve to the tune of "Big Bad John."

"BIG O"

Every morning in the studio, you could see him arrive,
He stood 6 foot 4 ... weighed two-twenty-five.
Kind of broad at the shoulder, and growing at the hip,
And everybody knew you didn't give no lip to Big "O".

Big "O" ... Big "O" ... Big Bad "O"

He came from a valley called the Kickapoo,
Where the cows all talk, and the milk is blue.
And early in the mornin' fore the sun would rise,
Them cows would be awaitin' with terror in their eyes for Big "O".

Big "O" ... Big "O" ... Big Bad "O"

It soon became apparent that he'd have to leave,
Cause pumping out utters didn't fill his need.
And so on to the radio he tried for fame,
And soon the whole county knew the name of Big "O".

Big "O" ... Big "O" ... Big Bad "O"

Small time radio was hardly the choice,
For the fair-haired man with the booming voice.
So to the big city he would try his hand,
At high-rise heaven in the promised land ... Big "O".

Big "O" ... Big "O" ... Big Bad "O"

He strode into GN, hands real steady,
They said, "You're on!"
He said, "I'm ready!"
Open that mike, shuck that corn,
Hello out there ... you're reborn with Big "O".

Big "O" ... Big "O" ... Big Bad "O"

His reputation grew and grew,
They liked that man from the Kickapoo.
From lookin' at fields, to trading beans,
From 3-piece suits to oversized jeans ... the Big "O".

Big "O" ... Big "O" ... Big Bad "O"

It's been 25 years since he gave up the yoke,
And started talkin' to just plain folk.
And through all that time there's few we'd expect,
Who haven't built up quite some respect for Big "O".

Big "O" ... Big "O" ... Big Bad "O"

So, Happy Anniversary from all of the team,
As broadcasters go, you are supreme.
Have a good time tonight, but keep it to beer,
'Cause if you're real good ... we're gonna buy you a Lear!

Big "O" ... Big "O" ... Big Bad "O"

WGN FARM REPORTING

"If you eat, you're involved in agriculture." That's one of my favorite sayings, and I have used it often to underline the importance of having farm and agribusiness reporting on WGN Radio. Still, the amount of farm broadcasting on WGN and on stations across the United States has steadily fallen over the years. Most of the managers of the big city radio stations have little knowledge of the food industry's importance to our daily lives. Their upbringing was probably in the city, and they don't really understand what happens west of the Tri-State Tollway. I can understand that, I really can, but over the years, especially the last 25 or so, I have tried to educate a significant number of general managers, program directors and radio sales people that WGN reaches a huge number of rural people who spend money on things like refrigerators and cars, just like city folks do. One of the first things Ward Quaal told me was, "We cannot look at WGN as a Chicago-only station, because we have farmers and ranchers out there who depend on us." I "outlived" nine WGN program managers who, when each of them first came on the job, said, "We have to get rid of the *Noon Show*, we lose the city listeners because of all that farm talk." Well, the program director who was in charge in 2006, finally got rid of the *Noon Show*. Agriculture made a lot of money for WGN. There were years where we would generate $4,000,000 of advertising revenue. The only rate on the rate card that was higher than the farm shows in the morning and noon was morning drive. And it got to the point where the market reports would have what we'd call adjacencies, and so farm advertisers would buy the minute next to the market reports throughout the day. It was very lucrative.

Things have changed. The Internet has diminished the importance of big stations like WGN in the lives of Americans. We don't generate that kind of money today because there are so many other avenues of advertising available to farm sponsors. And as the number of farmers declined, sponsors found even less reason to advertise. But we still maintain a significant amount of revenue, and the commitment to farmers, as of this writing in 2012, remains strong. We still do 16 reports a day, and Max Armstrong and I do an hour at five o'clock on Saturday mornings. So while the longer shows are gone, we do more market reports than we used to because of the volatility of the markets and how rapidly they can change during a trading session. And the content of the reports has changed. Until about 2003, the

WGN programmers wanted the farm market reports separated from the Wall Street reports. We finally convinced them that these markets are interrelated, so instead of doing Wall Street in one segment and Board of Trade and Mercantile Exchange in another, we do them together, because they are related. The price of crude oil, particularly, will relate to the price of corn and soybean oil that goes into the alternative fuels program — the ethanol and the biodiesel. Just as Wall Street is the center of the financial trading, LaSalle Street in Chicago is where the world's grain and livestock prices are set each day. It's a very big, very important market that we need to cover for Midwest farmers as well as investors.

WGN still has a large farm audience and I don't know of any audience that's more loyal. Day in and day out, farmers depend on what we do. When I came to Chicago, I felt my primary responsibility was to communicate to producers the weather, markets and whatever else they needed to make their daily decisions. But over the years, I came to understand that it's equally important to help the non-farm consumer in our audience — and that's the biggest part of our audience — understand what it takes to put food on their table. I have gone beyond being strictly a reporter. I've become an advocate for the people who produce livestock and crops. It's important that they have advocates because such a small number of people, around two percent of the U.S. population, are involved in producing our food. They need a voice and I'm proud to lend mine. I hear from producers over and over again, "Thank God you're here doing what you do," but I also hear from city people, "You know, without you, we wouldn't know why food prices are where they are, or we wouldn't know where the next county fair is going to be held," and that sort of thing. So I've come to see it as almost an equal responsibility — talking to farmers, talking for farmers.

Farmers have been getting better about telling their own stories, using all of the social media tools that marketers of any service or commodity use. But, they are often so busy doing the hard work of producing our food, the work they do so well, that they don't have enough time to go out and tell their story. That's one of the reasons I'm an advocate for them, because I have a voice. I've got a 50,000-watt transmitter, and a radio network of hundreds of stations. Max and I have a syndicated television show, *This Week in AgriBusiness* that is carried on TV stations across the Midwest, as well as on RFD-TV, a satellite channel aimed at rural folks. I suppose if farmers were 50 percent of the population instead of two percent, I wouldn't be as enamored with the idea that somebody needs to be an advocate for what they do, but I think I can help in many ways, including educating people who are trying to make laws and do other things that would make it difficult to farm.

That's one of my concerns, that we're passing rules and regulations in agricultural production today that ultimately could drive that production out of the U.S., where we'd have no control at all over what growing practices or humane practices are used with livestock. That's why I try to explain to the non-farm audience why we do what we do in agriculture and the fact that farmers are very careful and take good care of their animals and their land. I love covering the Environmental Stewardship Award winners each year, because those are positive stories about what farm families do in the way of conservation, in the way of no-tillage, in the way of wildlife habitat, and all of the other things they do that nobody knows about if we don't tell them. There's no other show out there like ours, so, yes, I'm an advocate and it's an honor to be one.

We covered some big stories over the years, both farm related and not. One of the biggest stories in my world of agriculture happened on a Saturday in July of 1972 when I walked into the office to do the *Country Fair* noon show. We still had the teletypes; Associated Press and United Press International with bells that sounded when there was a bulletin. When I arrived, the bells were ringing, so I walked over and watched as the teletype tapped out: "Secretary of Agriculture Earl Butz today announced a major sale of U.S. wheat to the Soviet Union." That turned American agriculture into a global agriculture, the beginning of the export and sale of American products overseas.

Another time, when I walked in early one morning and the teletype machine bells were dinging, I saw the bulletin that eight nurses had been found murdered in Chicago; the Richard Speck murders. But, easily the most memorable story happened when I was delivering the Midwest weather forecast on my noon hour show, *Country Fair.* It was an unseasonably warm late autumn day in Chicago, about 65 degrees and raining. As I was telling our listeners that a sharp change in the weather was coming with the next day's high temperature to be 20 degrees cooler, Gene Doretti, from the WGN newsroom, rushed into the studio and handed me a small strip of yellow paper that had been ripped from the teletype. It was a bulletin from United Press International and as I glanced at it while still reading the forecast, I thought, this is really a bad joke to play while I was on the air. When I looked through the studio glass into the control room and saw the expression on Gene's face as well as on engineer Bob Siebold's, I knew it was for real.

It was just three short sentences. I gave a time check: "Nineteen minutes before one o'clock, this is *Country Fair* from WGN Radio in Chicago." (pause) "Here is a bulletin from WGN News just handed me. President Kennedy has been shot and seriously wounded. Kennedy was shot just as his motorcade left downtown Dallas. A photographer on the scene says he saw blood on the President's head. I repeat, this bulletin from WGN News: President Kennedy has been shot and seriously wounded. Kennedy was shot just as his motorcade left downtown Dallas. A photographer on the scene said he saw blood on the President's head. This has been a bulletin from WGN News. WGN News will keep you posted on the latest developments as quickly as they are received." (pause) "In the meantime, we will continue with our *Country Fair* show. We'll have bulletins up to the minute as we receive them from the newsroom."

It was Friday, November 22, 1963. This was before we had cell phones, satellite television, computers and the Internet. News didn't travel as quickly as it does now and we relied upon UPI and AP to give us all of the news occurring outside Chicago. But that was all the information we had; three sentences stating that the President had been shot in Dallas.

So, what do I do now? It's the noon hour and the news director, the program director and the station manager were all at lunch, out of the building. At that time, it was required that a union musician be in the studio to play records, and he was out to lunch, so I couldn't play music. So I continued to deliver my normal program information, but every couple of minutes, I would repeat the bulletin. We finally located the record turner and at ten minutes before one, we went to music as we waited for more information. Just before 1:00 p.m., we received the bulletin that President Kennedy had died. At the same time, we were watching a television set that had been brought into the studio and we saw Walter Cronkite take off his horn-rimmed glasses and, choking back his tears, announce that President Kennedy was dead. I'll certainly never forget that day.

UPR95
BULLETIN
(DALLAS)---PRESIDENT KENNEDY IS DEAD.
A136PCD11/22

Within the WGN Farm Service Department, I had the opportunity to work with many people who became household names across the Midwest corn and soybean belts. As I mention a few of them, I'll start with the one who was with me the longest.

MAX ARMSTRONG

Max grew up on an Indiana farm, graduated from Purdue University and went to work for the Illinois Farm Bureau. That's where I first heard his voice. He was producing reports each week on a variety of farming issues that were put on tapes and sent to radio stations all over the state, including WGN. After listening to this young fellow, I turned to Lottie Kearns, my assistant in the Farm Department and said, "If I ever need a new colleague, I'm going to talk to this man."

In 1977, I wanted to hire Max, but he didn't seem that interested in coming to Chicago. Very similar to when Ward Quaal was trying to hire me in 1960, it took me over a week to convince Max to come to WGN. He had just bought a house in Towanda, Illinois where he was a volunteer fireman. Max loves to fight fires and later became a fire commissioner in Lisle, Illinois. I knew Max's boss, Bill Allen, and called Bill first before I even approached Max.

Bill said, "Well Orion, I'd hate to lose him, but an opportunity like this, he just can't turn down." But Max kept turning it down and finally, I called Bill and said, "I need help!"

Max told me that he had a day off and was at home. Around lunch-time, Bill Allen came rolling in with a couple of McDonald's hamburgers and said, "Max, we've got to talk." He told Max about the WGN offer, "You have to do this. It's a great opportunity that may not come around again."

Max and I worked together beautifully for 31 years at WGN and even though he left as a full-time employee of WGN in 2009 to join the Farm Progress companies, we still do the one-hour radio show together every Saturday morning at 5 a.m. We are also partners on *This Week In AgriBusiness,*which airs on television stations across the country and on DirecTV's RFD-TV, where we do our best each week to tell agriculture's story.

I'll step aside for a moment and let Max have a few words.

I was scared to death going to interview with Orion for the WGN job. And I don't recall it being quite as difficult to hire me as Orion remembers; I suppose I tried to play a little hard to get because we had just bought a home and I was enjoying riding around on the back of a firetruck as a volunteer firefighter. But I took it, obviously, and was on the job only a day and a half when Orion left town and I was on my own. It was "sink or swim. Figure it out, kid." Luckily, Lottie, our tremendous, very capable assistant for so many years, helped get me through it.

I was working for the Illinois Farm Bureau right after college and there were various events that I knew Orion and my predecessor, his former colleague, Bill Mason, couldn't get to. So, I would put together short reports and send them to the WGN recording line. They started using them on the air, which amounted to an on-air audition for me. In the summer of 1977, the Chicago Board of Trade had a media dinner one evening and those were splendid events. We were seated right down in the pits and the Board of Trade folks pulled out all the stops with after dinner brandy and cigars, right there in the pits. After the dinner, I went up to Orion and introduced myself as the guy who was filing reports from central Illinois and told him, "If you ever have an opening at WGN, I'd love to talk to you about it." Little did I know that four months later, I would have that opportunity. Bill Mason decided to leave WGN and Orion called my boss at the Illinois Farm Bureau, the late Bill Allen, and asked if it would be okay if he talked with me. At the age of 24, I was thrilled to death to go to work at WGN Radio. I was the youngest on the air by about seven years. I wasn't sure that I would last more than a few weeks but at least I'd be able to add it to my resume.

Working with Orion has had many benefits simply because you get painted with that brush of being his colleague. That's usually good, but there was an occasion — I want to say it was around 1990 when I was coming back from Canada — when I ran into a very interesting challenge because I was recognized as being his partner.

I had spoken to the Flax Growers of Western Canada in Winnipeg. My flight back to O'Hare was non-stop, but the plane was scheduled to go into a domestic terminal, so we had to clear U.S. Customs in Winnipeg. I'm standing in line and when it's my turn, an agent looked at my passport, looked up at my face, back to the passport, then up to my face a second time. He handed my passport back to me and pointed toward a room to the side. "Please bring your bag and join us in that room across there."

WGN Radio

As I headed for the room, and he followed me. I noticed him gesture to the other three agents to join him. They did, leaving dozens of people waiting in line.

So four customs agents followed me into the room, and I thought about making a joke, but it seemed to me that these guys were pretty serious about whatever was going on and it probably wasn't a good time to crack a joke. It was a little room, like one of those interrogation rooms you see on television shows, no windows, a table and a few chairs, very spartan. The other agents took turns looking at my passport then at me. I began thinking that this wasn't looking very good, that I might not be getting home that night.

As the door closed, the agent who had directed me to the room, started smiling and said, "You work with Orion right?!"

I said, "Well, yes, yes, I do."

"With Orion Samuelson, right?" he repeated.

Again, "Yes, I do."

The other agents chimed in, "Oh yeah, we're North Dakota farm boys. We've been watching him since we were kids. We just want to know what Orion is like."

I wound up being given the names and addresses of a dozen people in North Dakota who were to get autographed photos of Orion: Aunt Matilda... everybody in their families. Of course, I was more than happy to get them whatever they wanted. After they handed back my passport and as I headed for my gate, the agents told me to have a nice day and, "Tell Orion we said hello!"

JOHN ALMBURG

When I interviewed with Ward Quaal about becoming the Farm Director at WGN Radio and he outlined what kind of coverage he wanted me to provide, I indicated to him that it was more than a one-person job and in order to travel to the meetings and events we needed to cover, there would have to be two of us so that there would always be one person in the studio to cover all of the reports we provided during the day. He agreed, so one of my first jobs after arriving at WGN was to find someone to work with me.

I found John Almburg. John, from an Illinois farm, served in the U.S. Army and then was a meat inspector in Japan. Back home, he became a specialist with the U.S. Livestock and Meat Board in Chicago. He traveled the country doing television shows about cutting, curing and cooking meat. He had a good voice and served as a market reporter for the Chicago Union

WGN Radio

John Almburg on the air in the mid-'60s.

They travel 65,000 miles a year on business for you

In fact, their business is your business. They're farm directors Orion Samuelson and John Almburg of the big voice of Mid-America farming—WGN-Radio, Chicago.

In places like Washington, D.C.; Verona, Wisconsin; Clinton, Iowa; and Pecatonica, Illinois, they talk to the nation's most knowledgeable farming people. People like Department of Agriculture officials, farm organization leaders—and *you*.

You hear the results on their daily WGN broadcasts. Facts to help you make more money. Plus farm news, weather, on-the-scene market reports, reports from the eight major Midwest markets, and *lively* entertainment.

Get to know Orion and John at 720 on your radio dial. Stay with them throughout the day, every day, for complete farm coverage that comes close to home.

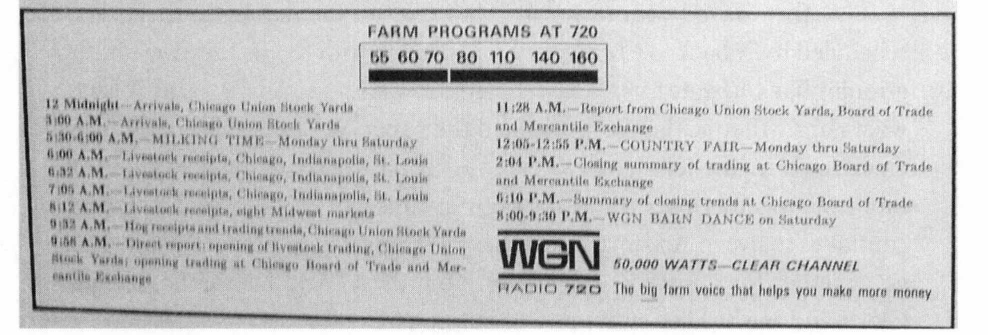

Stockyards and his reports were heard throughout the Midwest on WLS and other radio stations. I approached John and told him I'd really like to have him come on-board with me. He did, and we had a very good relationship and worked very well together. All of us who crossed his path over the years remember him for his great personality and sense of humor, as well as his passion for agriculture.

John left WGN in 1968 to devote his full attention to his first love: auctioneering and farm land. He started an auction and farm real estate company. Over 40 years, John conducted thousands of auctions across the Midwest and donated his auctioneering time and talent to numerous causes. John was a livestock auctioneer at dozens of 4-H auctions and because he had been at it so long, he had the opportunity to sell the livestock from families spanning three generations. At the time of his death in 2008 at the age of 78, John was still an active co-owner of Almburg Auctioneering along with his son Steve and grandson Andrew. He also enjoyed teaching aspiring auctioneers for over 30 years and was doing just that, teaching auctioneering to a student group in Fort Wayne, when he died of a heart attack.

After John moved on from WGN, we saw each other at meetings once in a while, but that was pretty much it, until the late '80s or '90s, when I already had a home in Arizona. My late wife, Judy, was very active in the Scottsdale Foundation for the Handicapped. There was an annual auction to raise money for the foundation and I auctioned it one year, but I'm not an auctioneer, so I thought, I'm going to get John Almburg. I called him out of the blue and said, "I'm giving you a ten-month warning. We'd like you and Donna to come to Arizona. You can stay with us and you can help us out with the auction." They agreed. The following year, in the spring, they came to Arizona and I did the emcee work for the banquet and John did the auction and it was very, very successful.

They stayed with us for five days and on the day when they were scheduled to fly back to Chicago, we had brunch with them. They were both grinning like Cheshire cats. I knew something was up and finally, I said, "Okay, what's up?" They said, "We just signed the papers on a house!" So, I'll take credit for bringing John and Donna to Arizona. We spent many wonderful moments together over the years, both in Chicago and Scottsdale. Donna continues to live in Arizona and we have a strong, friendly relationship and get together for dinner whenever we're out there. And Donna now helps Gloria and me by keeping an eye on our house.

BILL MASON

When John Almburg left WGN, I needed to find somebody else. Bill Mason had been an active broadcaster in agriculture, well known in the Chicago market. Bill was an Illinois farm boy, getting his degree in Agriculture Journalism from the U of I, then serving as farm service director of the university's Extension Service before going to work in Chicago radio. He was another WLS refugee, but also worked at WBBM and WCFL. Bill was boisterous and outgoing with a warm, down-home personality that listeners loved, but he had some very strong opinions. He and I knew each other well from both being members of the National Association of Farm Broadcasting and we didn't always see eye-to-eye on farm issues. At WGN, he would be working for me, a new path for him because he had always been the boss at WLS and WBBM. And he was certainly better known across the Midwest than I was when I arrived at WGN, because he had been on the air at those other stations. But, we agreed that it could work and he came to WGN. He stayed until the mid-'70s when he decided to go in a different direction that wasn't acceptable to WGN and so, with that, we parted company. It wasn't the friendliest parting, but later, at farm broadcasting conventions, we got along just fine.

Bill would often bring his sons to the station and to remotes, and one of them, Taylor, was bitten by "the bug." In 1990, Taylor Mason won

My former Farm Department partner Bill Mason, left, Bob Collins and me.

Ed McMahon's *Star Search* with his ventriloquism act, which started a very successful career. He has been touring with *Bill Gaither's Homecoming Tour* for several years.

Bill Mason passed away in 1997 at the age of 72.

LOTTIE KEARNS

A special salute goes to Lottie Kearns, who started as secretary to the WGN Radio Farm Department and ultimately became the *Noon Show* producer, particularly our on-the-road producer when we would take *Country Fair*, at first, and then the *Noon Show*, to the county and state fairs and the outdoor equipment and agriculture shows. She worked with us for 36 years.

Lottie had an interesting background. She was born in Germany in a refugee camp after World War II. When she and her parents moved to America, Lottie Americanized quickly, but her parents did not, and never did. They were so concerned about food, having gone through food shortages during the war in Europe, that they always had a three-month supply of lard and flour and other food items in their home. They didn't trust banks, again because of their war experiences, so they would stash money around the house. After her parents passed, Lottie found several areas where they had tucked money in the ceiling of the basement and other places where they thought it would be safe.

Lottie's father was a meat cutter and was quite thrilled when she became secretary to "Orient" Samuelson, as he referred to me. Every year when we did our broadcasts from the International Amphitheater, he made a point of coming out to visit us. He would take his hat off, shake my hand and tell me how proud he was of his daughter. Lottie had a wonderful personality that probably did more to sell WGN to our listeners than anything we did on the air at our remote broadcasts. Everybody got to know Lottie as we talked about her quite often on the air, but it was her sparkling personality that people were attracted to. And she was, and is, a very pretty lady who truly enjoyed meeting people.

It was a very sad day when the radio station management decided to do away with the *Noon Show* and with that, they said there was no longer any room for Lottie on our staff. Max Armstrong and I fought as hard as we could to keep her, but it had been determined by management that she wasn't going to be there any longer.

I still, in 2012, run into people who remember Lottie and ask how she's doing. She is doing well, but I know she misses the road trips and the great WGN listeners. Lottie was invaluable to me. She handled my calendar and booked my speaking engagements and had a tremendous work ethic. I couldn't have asked for more than she gave me and without her, I don't think I would have been able to accomplish all I did and keep the schedule I've kept over the years. We do get together for lunch now and then and catch up with each other. Just a great lady.

Lottie: One of my favorite memories of working with Orion was when we were driving to the Joliet Stockyards to do a "Noon Show" remote broadcast. Somehow we got a little lost (or as Orion said, "temporarily confused"). There was a police officer nearby, so we stopped him and Orion asked how to get to the Joliet Stockyards. As soon as Orion spoke, the police officer got all excited and said, "I know who you are!!!" I could see that Orion was flattered by the officer recognizing his voice, but the officer continued, "You're Milo Hamilton!" (Milo was the Cubs play-by-play announcer on WGN Radio at the time.) I laughed so hard and it was a moment I never let Orion forget. And even though he told me not to tell anyone, I couldn't help it. I've told just about everyone!

With my 36-year assistant, Lottie Kearns.

LYLE DEAN

When we started the *Tribune Radio Network*, several people helped make it work, including Lottie, Barb Atsaves Pabst, Max Armstrong, Jim Anthony and Lyle Dean. Lyle was instrumental in getting radio station affiliates for the TRN. He had become a legendary news anchor at WLS during its rock 'n' roll heyday, later anchored newscasts at WGN for 25 years. Lyle was also an occasional fill-in on market reports for Max and me. After Lyle retired in 2006, lo and behold, his brother, Steve Alexander, came on the scene as my fill-in, so the family tradition continues.

MOVING BACK TO THE TRIBUNE TOWER

In 1986, the first floor of the Tribune Tower became vacant. The decision was made to separate radio from television, give them separate identities and move radio back to the Tower from Bradley Place. I remember walking into the Tower shortly after that, just to see what our space was going to look like, and there was nothing, just concrete pillars. No walls anywhere, just wide open. Moving downtown, for most of us working at WGN Radio, was not appreciated because it added significantly to our commute and we had to pay for parking. But, it didn't take us long to acclimate to the new setting because they built a magnificent radio station. World class, state-of-the-art at the time. The main studio, A, was huge and I had friends come in from small town radio stations across the country and, in awe, tell me their entire radio stations would fit inside Studio A. I really did enjoy being back on the Avenue, and when the Showcase Studio opened, it added to the recognition of WGN Radio as a separate entity within the Tribune Company.

Not everyone liked working in the Showcase. Bob Collins and Spike O'Dell were never comfortable there and I don't think Wally would have been, either.

In 2012, we moved again, this time to make room for a restaurant. WGN's beautiful, new studios are on the seventh floor, but the Showcase Studio remains on Michigan Avenue.

Ole and Lena had an argument while they were driving down a country road. After a while they got tired of repeating themselves and neither wanted to back down, so they drove along in silence.

As they passed a barnyard of mules and pigs, Lena sarcastically asked, "Relatives of yours?"

"Yup," Ole replied without missing a beat. "In-laws."

With Gloria at my induction into the National Radio Hall of Fame in 2003.

Chapter Five

How Gloria Saved My Life

On a Monday night in 2004, I was in Nashville hosting a weekly RFD-TV live telecast and during the show felt the beginning of a sore throat. At dinner after the show, it was much worse and I left to go to bed as I had to catch an early flight back to Chicago. I called Gloria and told her I wasn't feeling well and was going to bed early. She had seen me on television and remarked that my face looked "puffy" and said she would get me in to see a doctor when I got home the next day. Meanwhile, the pain just kept getting rawer. It was unlike any sore throat I'd ever had, but a stubborn Norwegian doesn't go to the doctor for a sore throat.

Luckily, Gloria is "Southside Chicago Polish don't-mess-with-me tough," plus is somewhat of a worry wart with the additional trait of refusing to sit back and do nothing about whatever it is she's worried about.

Gloria: At first I thought the puffiness might have just been the TV cameras so after the show I waited for his call. He told me he had gone down for a bowl of soup and had a hard time swallowing it and that his throat was really sore. He called me a little later and could hardly talk. I called our new family doctor about 10 p.m. Dr. Alan Kanter had never met Orion and had only seen me once. I told him I thought Orion had strep and since he was out of town, could he prescribe some antibiotics that I could give Orion immediately when he got home. He was kind enough to do that. When Orion walked in the door about noon the next day I took one look and thought, oh my God, this man is sick! Of course, Orion said it was just a sore throat. But, I could tell that it was more than that. He was swollen and just didn't look like himself. My sister and her husband happened to be there and they also thought he looked sick. I gave him the antibiotic before he even took his coat off and put him to bed.

My appointment wasn't until four that afternoon, so I crawled into bed as the fatigue I had felt the evening before was getting worse along with the pain.

When we got to Dr. Kanter's office, he took one look inside my throat and knew it wasn't anything he'd ever seen and sent me to a throat specialist, Dr. Ari Taitz. He put a scope down my throat, looked for a few seconds, then called in another doctor to take a look. Dr. Taitz then ordered me to get to the Highland Park Hospital. "Don't stop at the Admissions Desk, don't stop in the Emergency Room, go directly to Intensive Care." When we walked into the Intensive Care Unit, doctors and nurses were waiting for me, started taking my clothes off and hooked me up to an I.V. I overheard one of the doctors ask if the "tracheotomy kit" was ready, which is something you never want to hear. Doctor after doctor peered into my throat and from their reactions, I could tell that they were seeing something that they hadn't seen very often, if at all. It was a severe case of the same illness thought to have killed George Washington, acute epiglottitis. A variety of bacteria was attacking my throat and I spent the next few days on an I.V. drip. I seemed to be getting better, but before sending me home, Dr. Taitz wanted to run the scope down my throat again.

"Orion, you aren't going home yet," he said. Instead, I was put on a stretcher, rolled into an ambulance and taken to Evanston Northwestern Hospital to be prepped immediately for surgery.

That's where I met Dr. Barry Wenig, the Chief of Otolaryngology and Head and Neck Surgery at Evanston Northwestern Hospital. He ran some tests and told me that my unusually sore throat was being caused by necrotizing fasciitis, more commonly called flesh-eating bacteria. Dr. Wenig said in his 30 years of practice, he had seen maybe five cases of flesh-eating bacteria in the throat. He said I needed to go into surgery immediately, but the more pressing problem was that I could barely breathe because my throat was swelling so quickly. Doctors had only a few minutes to perform an emergency tracheotomy, otherwise they say I would have suffocated.

When Dr. Wenig described how he was going to slit me from earlobe to earlobe and cut into my throat to snip out the tissue that was infected by the bacteria, I told him, "You can't cut me there, I make my living by talking." He looked me straight in the eye and said "Orion, in order to talk, you have to be alive."

"Well," I said, "since you put it that way, go ahead."

Gloria: Dr. Wenig came out of surgery after four hours and told me it had gone well. He explained how they had to remove the floor of Orion's throat and several muscles but, thankfully, they didn't have to touch his vocal cords; he would be able to talk. But there was a big gap in his throat and when they took off the dressings you could see everything. Dr. Wenig said they would cover the hole by rebuilding it. He would take a muscle from Orion's ribs and run it up under his collar bone into his throat to cover the area they opened up. He said that would be another long surgery, probably another four hours, because they had to shape the muscle to fill the gaping hole. But, two hours after he started, Dr. Wenig came into the waiting room with a grin on his face. He said, "The amazing thing was that the shape and size of the muscle we took from his ribs was exactly what we needed. It fit perfectly. All I had to do was stitch it in. Everything went beautifully."

After I'd come out of recovery and was lucid enough, the surgeon explained what he found when he opened me up. The tissue that was infected, actually killed, by the flesh-eating bacteria was a distinct yellow color, so it was a matter of just cutting it all out and stopping its rapid spread. Dr. Wenig started with the tissue that was perilously close to my vocal cords and worked outward.

I was in the I.C.U. for 21 days. I couldn't talk and had to be fed through a tube the entire time.

After leaving the hospital, I needed quite a bit more bed rest at home. My recovery was going well, the incisions on my throat were healing well. But one morning I could barely get out of bed and I was so weak, I could hardly walk. Gloria had gone to New York for a nephew's wedding, but my son David and his family were staying with me. They called 9-1-1 and an ambulance took me to Glenbrook Hospital where doctors found a blood clot between my heart and lungs. The staff stabilized me, then I got another ambulance ride to Evanston Northwestern Hospital. It was back to surgery, where they implanted a screen to catch pieces of the clot as they used blood-thinners to break it up.

As you'd expect, I pay more attention now when I get a sore throat because it's possible for the flesh-eating bacteria to return. But as for how it got there in the first place, I have no idea. Doctors asked if I'd been to a foreign country and I had, but that was 18 months earlier. There's no specific cause of it and it more often than not leads to death if it isn't treated quickly enough. Dr. Wenig told me later, "You are one lucky man, Orion. Those bacteria were spreading so quickly, you were less than an hour from being dead." If Gloria hadn't gotten me started on antibiotics as soon as I got home from Nashville, she likely would have become a widow.

Gloria: He was a sick puppy. If it wasn't for the first doctor, Dr. Kanter, prescribing antibiotics and stemming it a little we would have lost him. I do thank him for being so kind, quick acting and believing what I was telling him. Maybe it was my desperation! He was so instrumental in getting the next two doctors involved and thus saving his life.

We thank God for the three doctors and their staffs who made all the right decisions at the right time. Had Dr. Kanter not prescribed the antibiotics for somebody he'd never met, had Dr. Taitz not acted with expediency, and had Dr. Wenig not been the skillful (not to mention compassionate) surgeon, I wouldn't be writing this. Gloria calls those three doctors her personal heroes and I couldn't agree more, but Gloria is my heroine. She saved my life.

The ordeal left a lasting impression on me. I tend to squeeze hands a little firmer, linger with our friends a little longer and hug my grandkids a little tighter... and listen to my wife!

Ole had been in a bad crash on the country road near his home. He decided to sue for damages. In court, the insurance company's attorney was trying to prove that Ole was faking his injuries.

"Didn't you tell the state patrolman at the scene of the accident that you were, and I quote, 'Oh, ya, I'm doin' okay officer. I'm fine?'"

"Vell, here's vaht happened. I had just loaded Bessie, my favorite mule,

into the trailer and vas driving her down da highway ven dis huge semi ran the stop sign and smacked my truck right in the side. I vas thrown into one ditch and Bessie vas thrown into the other. I vas hurting real bad and didn't vant to move. I could hear Bessie moaning and groaning and knew she was in terrible shape.

"Just then, a Highway Patrolman drove up. He could hear Bessie moaning and groaning so he went over to her. After he looked at her and saw how bad off she was, he took out his gun and shot her between the eyes. Then the patrolman walked across the road, gun still in hand, looked at me and said, 'How are you feeling?' Now, vaht vould YOU say?"

The one and only Paul Harvey, 2003.

Chapter Six

Paul Harvey and the National Radio Hall of Fame

Of the many honors for which I'm very grateful, one of the most special was being inducted into the National Radio Hall of Fame. Imagine being included in an elite group that featured such legends as Fred Allen, Arthur Godfrey, Bob Hope, Bing Crosby, Gene Autry and Paul Harvey. It's very humbling and, frankly, when Bruce DuMont, the founder of the Museum of Broadcast Communications, told me I had been nominated I was very flattered, but didn't think it would go any further. When I saw the others nominated in my category, big radio names from New York and other major cities, I didn't think I had a chance. So I told myself it was a great honor to be nominated and expected nothing further. When Bruce called with the news that I had been chosen for induction, I was shocked.

Each inductee can choose who they want to do their introduction. My first call was to Ward Quaal, the man who hired me at WGN in 1960. Unfortunately, Ward had a long-standing commitment to an event on the same day in Los Angeles and wasn't able to get out of it. So, I called Bruce and said, "Do you think there's a chance that Paul would introduce me?" Bruce said Paul had done several inductee introductions and after the most recent one, he told Bruce that he thought he'd done his share and didn't want to do any others. "But, I'll call and ask," said Bruce. Ten minutes later, he called back. "Paul said he would be delighted to do it." What an honor for me. Here is what Paul Harvey said (as only Paul could say it) to the assembled crowd on Saturday evening, November 8, 2003 at the Chicago Cultural Center:

"Good evening, Americans. I'm told I have ninety seconds. ABC complains that Paul Harvey's pregnant pauses sometimes last longer than that. But, here goes. Farmland and Ranchland, USA, no longer have the numbers on which our media thrive. So, that vital, crucial, absolutely indispensable segment of our nation's population needs a big voice. Orion Samuelson's is a biiiiiiiiig voice!

Orion Samuelson grew up with his feet in a furrow. Orion could have been our nation's Secretary of Agriculture under any of three presidents. But, he was too smart for that. Instead, for eight presidents over more than four decades, the Big O of WGN Radio in Chicago and on hundreds of stations every day feeds the men and women who feed us the information they'd not get any other way. As attested by the highest award of the American Farm Bureau and the highest honor of his home state and the myriad recognitions in between our industry, broadcasting, has achieved its highest purpose in serving the public interest, convenience and necessity in farm land through the diligent dedication of one giant of a man. In our radio audience, for the people listening tonight in our radio audience, let's try to visualize one guy representing two million. He's made of bent nails and rusty horseshoes and barbed wire and held together by calluses. With his wife, he's hunkered over a tabletop radio in the kitchen. He's hurting from tractor back. But, he's up way past bedtime because he's heard that his best friend is being honored tonight. Well, by golly, he's right. And, I want you to note that even the moon above is hiding its face at this moment, rather than to distract from the significance of this introduction of Orion Samuelson!"

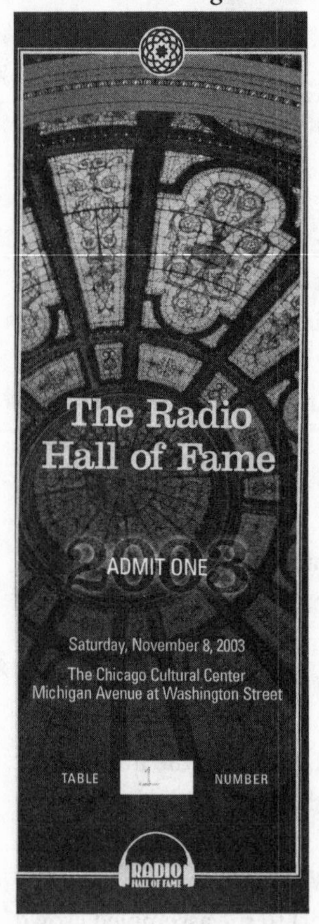

That night, I sat at the table before the presentation with Paul and his wife Angel and my wife Gloria. During the meal, Paul reached into the inner pocket of his tuxedo jacket, pulled out his 90-second script, read it, and tucked it

If the child's name is incorrect, return this card to the State Bureau of Vital Statistics with the correct information. This is very important.

During the latter half of its first year have your child protected against diphtheria. Over one-half of all diphtheria deaths occur under school age.

Bulletins on the care and feeding of infants will be sent free to any citizen of the state upon request.

MADISON SEP 6 2 7 PM 1934 WIS.

THIS SIDE OF CARD IS FOR ADDRESS

ONE CENT 1 JEFFERSON 1

Sidney O Samuelson
Ontario,
Wis.

This is what Wisconsin birth certificates looked like in 1934.

The World of Tomorrow is in the Hands of the Children of Today

WISCONSIN
STATE BOARD
OF HEALTH
MADISON

Certificate of Birth Registration

This is to Certify *that a registered certificate of the birth of your child has been filed and is now carefully preserved in the Official Records of the State of Wisconsin in the State Board of Health office at Madison.*

Name Orion Clifford Samuelson

Maiden Name of Mother Mona Paulson

Birth Place of Child Whitestown, Vernon

Date of Birth Mar. 31 1934

C. A. HARPER,
State Registrar of Vital Statistics

PRESERVE THIS RECORD

My dad as a boy on the Samuelson farm, circa 1910. From left: Grandmother Johanna (Jennie), Grandfather Ole, Uncle Harry and my father, Sidney (Sam).

On the farm with Mom, Dad and Sport, 1937.

BACK ROW: *Grandpa Ole Samuelson, Dad, Uncle Harry.*
FRONT ROW: *Aunt Madeline, Grandma Jennie, Aunt Myrtle.*
June, 1924

My mother, Mona Paulson, at her Confirmation.

The Samuelson barnyard and dairy herd, circa 1915.

The Samuelson farmhouse, my birthplace.

Counterclockwise from top left: Dad (Sam), Mom (Mona), me, and relatives from Mom's side of the family, Aunt Clem Paulson, her husband, Uncle Norman Paulson and Uncle Bernt Paulson.

Mom and one-year-old me on the farm, 1935.

My first farm implement, 1936.

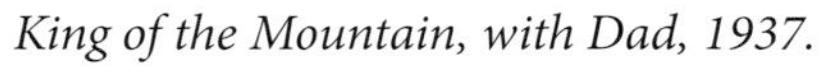

King of the Mountain, with Dad, 1937.

Enjoying the snow with Sport.

In the garden at four years of age, 1938.

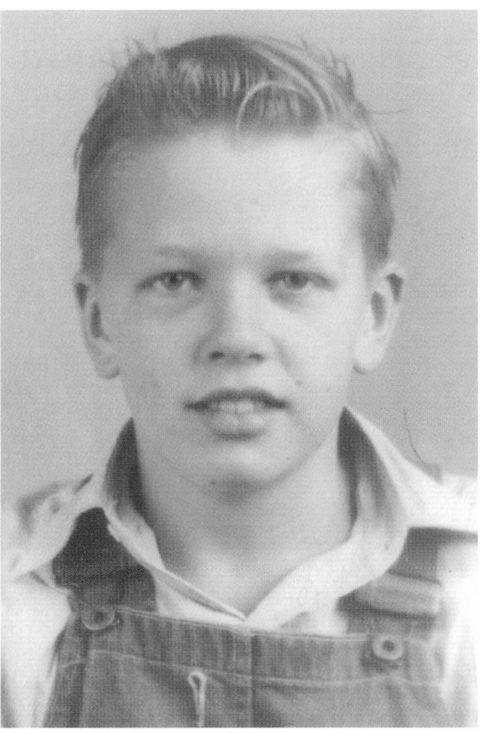

About 10 years old, wearing the standard Wisconsin farm boy overalls.

Age 12.

I vividly remember the day in 1939 when the tractor dealer delivered our shiny, new Farmall F-20. Like most old farm machinery, it became obsolete and was left to rust. On June 9, 2003, Max Armstrong and the International Harvester Collector's Club #10 of Central Illinois surprised me with a fully restored Samuelson F-20. BELOW: *My son David, grandson Matthew, Max and me.* OPPOSITE PAGE: *The F-20 looks better than brand new, but the driver has a few hours on him.*

FARMALL

My sister, Norma, married a local boy, Norman Haakenstad, on Halloween in 1959.
Along with five of their own children, they cared for over 130 foster children.
Norman was 78 when he passed away in January of 2012.

WGN Radio

Among the many perks of my job was meeting celebrities. Some, like Roy Rogers, were boyhood idols. This was at the International Livestock Exposition in Chicago, probably around 1970.

ABOVE: *With John Forsythe.* BELOW: *Miss America 1969, Judith Anne Ford of Belvidere, Illinois and band leader Mitch Miller.*

WGN Radio

A group of farm broadcasters crowd around Dinah Shore. FROM LEFT: *Me, Herb Plambeck, Jack Crowner, Dinah, Bill Mason, Bill Arford and Hugh Ferguson.*
BELOW: *With baseball Hall of Fame pitcher and broadcaster Dizzy Dean.*

Leroy "The Auctioneer" Van Dyke, his wife Gladys, Gloria and me in May of 2000. The Van Dykes, along with their son Ben, surprised me that spring by naming a newborn mule after me, shown below as a yearling: Sam-Mule-Son.

For a number of years, I've had the pleasure of co-hosting RFD-TV's live coverage of the Tournament of Roses Parade in Pasadena, California. My co-host is Pam Minick, World Champion Calf Roper and former Miss Rodeo America.

Chicagoan Clara Peller became famous in 1984 with her shout of "Where's the beef?!" in Wendy's commercials. We're shown at the Blossom Time Festival in Benton Harbor, Michigan.

back into his pocket. He did that three times. That truly impressed me. A man who has been in broadcasting as long as he had and was probably the most respected broadcaster in the world, and he still feels that it's important to rehearse and be prepared. He did a magnificent job, of course, and that closing line about the moon is one that I'll never forget.

Paul was "the most listened-to man in America." He was a masterful broadcaster. His use of the pause was just magnificent. Paul would pause and you just waited to see where he was going with it. He knew how to get your attention and how to keep it. I didn't get to know Paul until the last 15 or 20 years of his life. It started with a telephone call from Paul and he told me he was a listener and in many ways, we shared similar backgrounds with the rural life that we both came from and loved. He said, "Orion, occasionally I do an agricultural story that I may want to verify. Is it okay if I give you a call?" Well, of course I said, "Anytime! I'll be happy to share with you whatever I can that will be helpful to you."

Whenever Paul would call me, the conversation started like this:

"Farm Department, Orion speaking," I'd say.

"Hello American!" Paul would respond with that same energy and intonation he used when he started his broadcasts.

In the August before he died, he called me and said, "Orion, what can you tell me about sweet corn?"

"Sweet corn?"

"Yes, I am hearing about this Mirai sweet corn on your radio station and I'm curious about how it can be sweeter and tastier than regular sweet corn. Couldn't we accomplish the same thing by simply soaking and boiling regular sweet corn in sugar water?"

"Well, Paul," I chuckled, "I don't think that would work. I think you have to let Mother Nature do it."

But I told him, "I think you should talk to the people up at Harvard, Illinois at Twin Garden Farms who will be happy to tell you about their variety of sweet corn. It's a fascinating story because most of the seed they produce is shipped to Japan where people love the Mirai sweet corn."

I gave him the phone number for Gary Pack, whose grandfather started Twin Garden Farms. Later that afternoon, I got a phone call from Gary.

"Orion, I've been on the phone for the last hour and a half with Paul Harvey!" Gary thought it was someone playing a joke on him at first, but it was Paul and he invited Gary to visit with him in his office at Michigan and Wacker Drive in downtown Chicago, which he did. Gary, then, invited Paul to visit the farm, which Paul did. Paul had his driver take him out to Harvard and he spent an entire afternoon on the farm learning all about sweet corn. That's one of the things that impressed me about Paul, his curiosity.

Paul was just a great human being and what an honor it was for him to mention me on the air from time to time when I was given an award for one thing or another that was important to me. He would find time to mention it on his program and each time he did, I'd hear about it from a number of people across the country. And he was very gracious in everything he did, a true gentleman.

A few years before he died, I was sitting in his office before one of his broadcasts and he was typing his script on an old typewriter — this was well into the computer age, but he still used his typewriter — and then he began doing the vocal exercises he did each time before he went on the air. They were just gibberish... he'd be vibrating his lips and shaking his head and strange sounds came out of his throat. But when the microphone came on, it was magic.

In the fall of the year before he died, I called him and said, "Paul, you and I have rural backgrounds and you spend time on your farm in Missouri. I'd love to come over and sit down with you, turn the recorder on and have two rural kids who could never have dreamed what would happen in their careers just visit."

"Well, Orion," he said, "I'm not ready to do that, but we should have lunch first."

Paul had a passion for communities in rural America. During his speaking tour days he would often schedule appearances in small towns to deliver an address to a graduating high school class or a farm organization convention. And agriculture liked Paul Harvey. He scored his strongest ratings in the Heartland, as he called it. Companies like Archer Daniels Midland and farm organizations like the American Farm Bureau Federation bought time on the Paul Harvey News and Comment to carry their message to the 24,000,000 people who listened to Paul every day.

While some people were bothered by the conservative stance of Paul Harvey, I embraced it. But we did have one area of disagreement: his stance on animal rights and the admiration expressed on his broadcast for PETA (People for the Ethical Treatment of Animals). Often, after he did a story on

PETA, we would talk on the phone and I would tell him he was wrong. He didn't agree, but we disagreed civilly because that was Paul Harvey's way.

In 2007 at the wake for his wife of 67 years, Angel, he called me aside. "Orion, sit down and let's talk." I did and we talked about change of ownership at the Tribune and some of the things in the broadcast industry

ABOVE: *Paul Harvey watches as I accept my induction into the National Radio Hall of Fame on November 8, 2003.*
BELOW: *Bruce DuMont, founder of the Museum of Broadcast Communications.*

that bothered him, things that bothered me, too. For example, topics we felt should be in private conversation and not aired on the public airwaves, and the growing use of swear words on the air. We both felt if you couldn't make your point without using four-letter words, you were a poor communicator. I still feel that way. We talked for probably fifteen or twenty minutes and as we ended, I said, "Paul, we've got to have that lunch." We never got around to it, partly because we were both too busy. Paul never retired. He said, "Retiring is just practicing to be dead. That doesn't take any practice." He was still broadcasting the week before he died in 2009 at the age of 90.

Paul was a truly outstanding gentleman, broadcaster and friend who had respect for everyone and their views. That is why, for years, I have said, "When I grow up, I want to be like Paul Harvey."

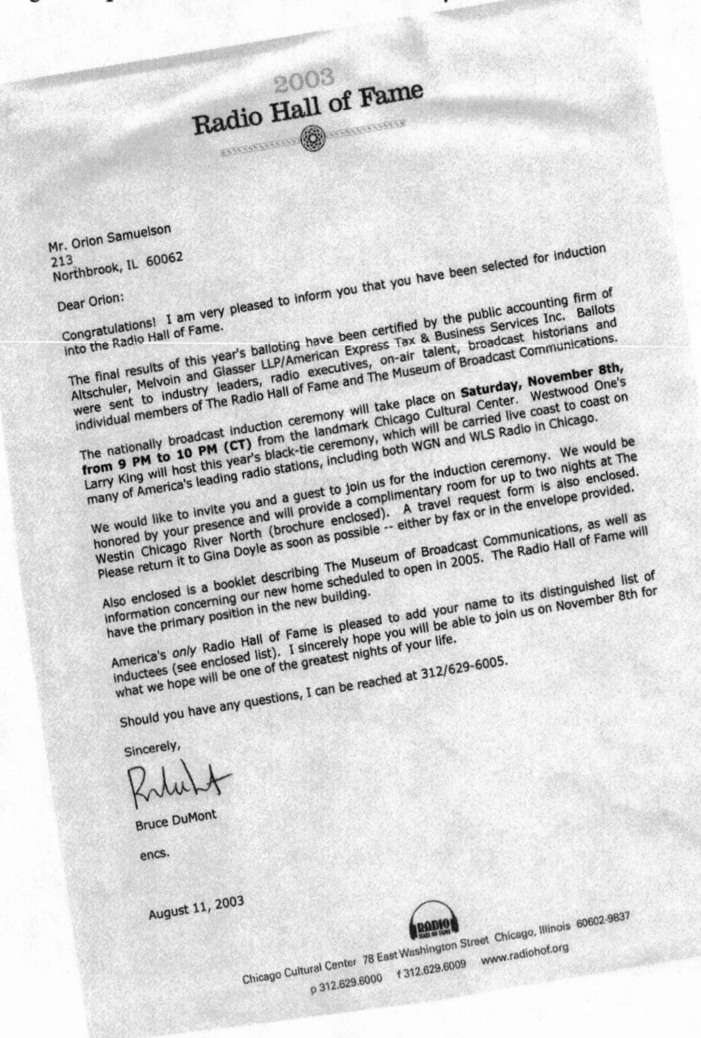

2003
Radio Hall of Fame

Mr. Orion Samuelson
213
Northbrook, IL 60062

Dear Orion:

Congratulations! I am very pleased to inform you that you have been selected for induction into the Radio Hall of Fame.

The final results of this year's balloting have been certified by the public accounting firm of Altschuler, Melvoin and Glasser LLP/American Express Tax & Business Services Inc. Ballots were sent to industry leaders, radio executives, on-air talent, broadcast historians and individual members of The Radio Hall of Fame and The Museum of Broadcast Communications.

The nationally broadcast induction ceremony will take place on **Saturday, November 8th, from 9 PM to 10 PM (CT)** from the landmark Chicago Cultural Center. Westwood One's Larry King will host this year's black-tie ceremony, which will be carried live coast to coast on many of America's leading radio stations, including both WGN and WLS Radio in Chicago.

We would like to invite you and a guest to join us for the induction ceremony. We would be honored by your presence and will provide a complimentary room for up to two nights at The Westin Chicago River North (brochure enclosed). A travel request form is also enclosed. Please return it to Gina Doyle as soon as possible -- either by fax or in the envelope provided.

Also enclosed is a booklet describing The Museum of Broadcast Communications, as well as information concerning our new home scheduled to open in 2005. The Radio Hall of Fame will have the primary position in the new building.

America's *only* Radio Hall of Fame is pleased to add your name to its distinguished list of inductees (see enclosed list). I sincerely hope you will be able to join us on November 8th for what we hope will be one of the greatest nights of your life.

Should you have any questions, I can be reached at 312/629-6005.

Sincerely,

Bruce DuMont

encs.

August 11, 2003

Chicago Cultural Center 78 East Washington Street Chicago, Illinois 60602-9837 www.radiohof.org
p 312.629.6000 f 312.629.6009

Ole had been courting Lena for quite some time and Lena was beginning to think that Ole would never ask her to marry him.

One evening, they went for Chinese food and as they looked over the menu, Ole said, "Lena, do ya vant your rice fried or boiled?"

Lena looked up from her menu and said, "I vould like my rice trown, Ole! And da sooner da better!"

In Cuba, sampling the famous cigars with Governor George Ryan and a couple of our hosts, 1999.

Chapter Seven

World Travels

Often, I'm asked if I have a favorite hobby. If Gloria were asked, I'm sure her answer would be, "His work is his hobby," and she would be right, because I do love my job. But, if I was writing my job description, I'd have to include, "ability and desire to travel the world." Travel is the best educational opportunity any of us can have and, in my case, my job has presented me with some wonderful opportunities. When I was crouched at cows' udders before dawn on ice-cold Wisconsin mornings, the thought of getting on an airliner was as foreign as the 43 nations to which I eventually traveled. Broadcasting opened up my world in ways I could never have imagined.

Curiosity has always been one of my dominant traits, and that thirst for wanting to know how farmers in other countries stacked up against ours, what kinds of crops they raised and how they harvested and processed them drove me to grab every opportunity to find out. Below, you'll read about a few of the far-flung places my traveling microphone landed. I'll begin with the land of my ancestors.

NORWAY

Gloria and I led six agricultural tours to Norway. The last trip, in 2009, was made up of 40 farmers and their spouses from 12 states: New York, Illinois, Wyoming, Iowa, Maryland, Tennessee, Minnesota, South and

My Norwegian cousin and tour guide, Carroll Juven, left, with our tour bus driver, Jan Erik Nybakken.

North Dakota, Texas, Kentucky and Utah. While we got acquainted with our new friends, we also met Norwegian farmers and their families on nine different farms in southern Norway. They are way ahead of us in embracing agri-tourism and at each visit we enjoyed lunch or dinner in the farm homes prepared and served by the family members.

A tour of the farms generally included a family museum because preserving the past is very important to these people, as it is to me. As I wrote in an earlier chapter about my family, my four grandparents grew up in Norway and came to America as young people. They settled with other Norwegians in western Wisconsin. On my first trip to Norway, I found part of my roots in a small clearing in a thick forest in a narrow valley near the town of Tretten. I stood in what was left of the stone foundation of the home where my grandmother Jenny was born in 1864. It turned out to be a much more emotional moment than I could have imagined as I looked around that clearing and realized that a part of me started there.

It also helped me understand what difficult choices had to be made by many parents in Norway and other European countries who hoped to give their children a better life by breaking up their families and sending the children across the ocean to live with other relatives in the U.S. What seems unthinkable in modern day America was common between 1850 and 1925 when 850,000 Norwegians made the often perilous journey. The visit to my roots helped me understand a great deal more about my family and I urge you, if at all possible, to search for your roots and take the opportunity to travel. Books and photos are nice, but being there and seeing it with your own eyes will be one of the great learning experiences of your life.

Before we left on that 2009 trip, I received an e-mail from a long-time Norwegian friend of mine saying, "Welcome to our World." He was not welcoming me to Norway, per se, but he was welcoming me to the world of socialism. Norway is one of the most socialistic countries in the world. As a matter of fact, their citizens joke that the government takes care of them from "womb to tomb." And my Norwegian friend said in his e-mail, "It looks like America is coming our way."

We may have slipped in that direction in recent years, but I doubt that we'll go all the way to "from womb to tomb." But this concern is nothing new. Let me share with you words from a speech delivered by Dick McGuire, President of the New York Farm Bureau. Dick said, "The congressional pastime of throwing money at problems to make them go away has become a national sport. There's another more frightening dimension of losing control of deficits and accelerated spending. Government itself tries to impose more and more control and with every increase of government, there is a corresponding reduction in individual freedom. We must regain control of government expenditures. First, we elect people that believe in balancing the budget. Second, we support a constitutional amendment requiring it yearly, by restricting spending to a realistic percent of gross national product. Between free enterprise and outright socialism is the condition toward which the United States is now gravitating... a constant increase of governmental interference in business where it makes more and more decisions, owns more and more property and takes more and more profit."

That speech was delivered in 1975.

The U.S. farmers who joined us on our tours of Norway farms were always impressed by the resourcefulness of the Norski farmers. One of the biggest challenges is that there isn't much land that's tillable. Norway has about 2.1 million acres of farmland. The State of Illinois has about 27 million acres that is farmed. Illinois farmland is mostly flat as far as the eye can see. Norwegian farming is done up and down hillsides and wherever flat areas can be cleared out in narrow valleys. What may surprise you is that the growing seasons aren't that much different. In Illinois, the number of frost-free days ranges between 190 and 200. In southern Norway, it's from 177 to 210 days and Norwegian summer days have more daylight hours than ours in the Midwest. But most of the land where the warmer weather is, is along the coastlines and there isn't very much of it that can be farmed.

I do give my Norwegian cousins credit for thinking ahead. In 2009, they opened the Svalbard Global Seed Vault, which is a secure seed bank tunneled deep into a mountain on an island about 800 miles from the North Pole. Hundreds of thousands of seeds from around the world are stored there at zero degrees Fahrenheit. The Norwegians built the seed vault as insurance in case seeds in other gene banks are lost or mismanaged, or there's some sort of catastrophic regional or global crisis.

THE LONGEST TRIP

In March of 1974, the Secretary of Agriculture's office called and said, "Secretary Butz would like you to be the radio/TV observer on a 23-day trade mission to Asia." Well, that is a major amount of time to be away and my expenses would have to be covered by my employer and not the USDA. Luckily, Ward Quaal was the president of WGN at the time and when I told him about the invitation, he didn't hesitate. "Orion, it is a great honor to be invited and WGN will foot the bill." So we went to work immediately to cover everything that needed to be covered. Bill Mason was my partner at the time and he had to clear his schedule to be available from April 2nd to April 25th to cover all the WGN Radio and television broadcasts. So, a very excited Big O prepared for the longest time away from the station and what would be my longest foreign trip.

Tuesday, April 2nd we departed Los Angeles on a U.S. government plane. It was a converted KC-135 Tanker which is the military version of the Boeing 707 passenger plane, and had been configured for passengers. There were three media people: a magazine writer from St. Paul, a farm newspaper editor from Kansas City and myself. Also, Secretary Butz, his wife Mary, some top-level USDA officials and two full flight crews in addition to a crew of Air Force personnel who prepared meals and served us. We flew to Honolulu and landed at Hickam Field where we were greeted by Admiral Noel Gayler, the Pearl Harbor-based CINCPAC, the Commander-in-Chief Pacific Command. He was a living Naval legend, having collected an unprecedented three Navy Crosses in only four months as a pilot during World War II. As Pacific Command commander, just one year after our visit, he would oversee the fall of Saigon and the evacuation of 1,500 Americans and some 100,000 Vietnamese refugees. We went to the admiral's beautiful, 20-room home and spent a delightful three hours getting acquainted with him and his wife Connie. We learned that Admiral Gayler's father, Naval

President Lee of the Republic of China, presenting me with the International Communicator of the Year award.

Captain Ernest Gayler, had built that house in 1919. And then I learned that Mrs. Gayler grew up on a farm in Iowa. Her parents were cattle farmers who watched our weekly show, *U.S. Farm Report*. They happened to be visiting at the time and we had a nice visit with two cattle people from Iowa. Noel had his own "cattle story." When he was stationed at Norfolk, Virginia in the early '50s, he flew an experimental fighter jet to Denver and back non-stop. Because it was a demonstration of how the jet could be used in a low-level attack, he never flew higher than 200 feet. He said there were a lot of very surprised cattle along the way. Noel was born on Christmas Day, thus the name. In 1975, we were vacationing in Hawaii for the holidays and Noel invited us to come to his home for the dual celebration.

From there it was off to Guam and we landed at the very busy military base there. That was during the waning days of the Vietnam War and the B-52 bomb missions originated in Guam. I sat in the cockpit of a big B-52 and even when the plane was on the ground, it was pretty high. We had to fly around Vietnam because of the war and landed in Bangkok, Thailand on April 3rd about 10 p.m. We were greeted by the prime minister, a military band and a great ceremony. Earl Butz was the highest-ranking U.S. official to visit Thailand and they really put on a show for the three days we were there. I vividly recall the wild ride to the hotel from the airport. We had a military escort in front, a military escort in back and seven cars in between. The media people were in the last of the seven cars. Even that late at night, the streets were busy but that didn't slow down our motorcade. We were passing when we shouldn't have been passing, horns blowing constantly and the military car in the lead telling us to go faster, keep up and the military vehicle behind us right on our bumper. Luckily, we arrived at our hotel in one piece. One floor of the hotel had been cordoned off for the Butz party and when I was escorted to my room, my name had been carved in a wood plaque on the door! Secretary Butz dropped by every room to make sure everything was okay and working fine. The USDA staff had prepared a large notebook giving all the information we needed, from itinerary and location of meetings that the media was allowed to cover and those that were just for the Secretary and top Thai officials. It also contained do's and don'ts for foreigners, customs of the culture, how to get medical help, how to get in touch with your family back home if it was an emergency, etc. It was a thorough guidebook with everything we needed to know to make the most of our trip and stay out of trouble. Some of the interesting cultural customs:

- If you are walking on the sidewalk and you are approached by monks wearing orange robes, make sure you don't touch them or their robes. In fact, it advised that you cross the street and walk on the other side if you see the monks approaching.
- Another no-no was to never pat a child on the top of the head because the Thais felt that would interfere with bad spirits that were going out or good spirits that were coming in.
- If you are sitting when someone speaks to you, then you need to rise to talk to them.
- Never point your toes directly at the person you are talking to.

We left Bangkok on April 6th and flew to Hong Kong, stayed there overnight as well as the next day. From there, we flew to Manila. We met President Ferdinand Marcos, spent three days and left on the 11th of April. I must tell you about April 9th because it is a never to be forgotten day.

At the time we visited, April 9th was when the Filipinos commemorated Bataan Day. It was on April 9, 1942 that 12,000 American troops and five times that many Filipinos surrendered to the Japanese at the tip of the Bataan Peninsula, which juts into Manila Bay. For nearly five months, the troops had fought ferociously against overwhelming odds until they ran out of food, medical supplies and ammunition. As prisoners of war, they and thousands of Filipinos were marched as far as 55 miles to a train where they were transported to a camp run by the Japanese army. This grueling series of marches is now known as the Bataan Death March. It was the largest single surrender of American troops in the history of our country and the marches were among the darkest moments for our troops.

We were taken across the harbor from Manila to Mount Bataan where the Filipinos built a large, outdoor chapel with marble columns and a marble altar. Secretary Butz and his wife went across the harbor on the presidential yacht. The rest of us were in another, less luxurious boat and then took a bus to the memorial at the top of the mountain. From that hill you can see Corregidor, where one of the most horrific battles between the Americans and Japanese had taken place. I can still see President Marcos' face as he spoke to us at the Bataan Memorial. He had been in the battle and the Death March. As he told us his memories, he never once looked at those of us seated on the marble benches. His gaze was over our heads and with a distant eye said he never thought he would survive the brutal march and be standing there with us. It was a never to be forgotten day.

Also in the Philippines, I got to visit the International Rice Research Institute where some of Norman Borlaug's Nobel Prize-winning research had been put into practice. A lot of the work he had done in Mexico and India had been shared with the Research Institute. We also had an opportunity to visit a Filipino farmer who had 2 ½ acres of rice. Generally, rice was planted by hand and harvested by hand, but he had a rice variety that he could harvest with a new rice harvest machine. It was like an oversized lawn mower. I did get to interview the farmer and in the course of the interview he said that he had read about farming in America and had seen photographs of the big farms and big machinery. He said, "I would really like to spend just one day walking behind an American farmer to see what it is like to have 200 or 2000 acres instead of two, like I have." I kept his name and how to reach him and when I came back to America I worked with some friends to raise money to bring him over. But when I contacted him, he said, "Oh, no, I wouldn't know how to travel that far."

After Manila, it was off to Taipei. 1974 was an interesting time to be there because Taiwan still had rockets pointed at Red China, only 112 miles across the Formosa Strait, and rockets were pointed back at Taiwan. Taiwan insisted it was an independent nation and not part of mainland Communist China. It was a rather tense time. But I learned a lesson in Chinese culture. On the day we arrived I was sitting at a luncheon table with two Americans and six Taiwanese. Two of the Taiwanese were ranking officers in the Republic of China army and they kept saying, "We're going to retake the mainland." I listened to them tell us that a few times and finally said, "I don't mean to embarrass you but I have to ask; there are 800 million people on the mainland and there are 20 million here on the island. I don't understand how 20 million will overcome 800 million." One of the officers looked at me and said, "In Chinese history, dynasties rise and fall. They may last 600 or 800 years but eventually they will fall." I said, "Oh, you're not talking next year or ten years from now. I understand!" That was my lesson on the Chinese attitude towards history.

It was on to Tokyo where we had the opportunity to spend time in some food processing areas. Two years earlier, President Nixon placed an embargo on soybean shipments to foreign countries. At that time Japan was our biggest buyer of soybeans and soybean oil and they used it for human food. That angered the Japanese and resulted in them developing soybean production in South America so they would have another soybean source and not have to depend on us alone for their supply. It was a costly move by Nixon because it ultimately made Brazil a major competitor to U.S. farmers. Our itinerary included a visit to a supermarket in Tokyo and we had gotten quite a bit of press during our trip. We had been in newspapers and on Japanese television and, to this day, I don't know if this was set up or if it just happened. When I was walking down one of the aisles looking at prices a young Japanese lady with a baby approached me with a shopping cart and asked if I was with the secretary's group. I said yes and she said, "I just want to tell you that when you stopped selling soybeans to my country you took food out of my child's mouth. That scares us very much and I hope

ABOVE: *Two of the Pearl Harbor survivors attending the final reunion in 2002.*
BELOW: *After a flyover at the 65th commemoration, a ship steamed by, all hands on deck and saluting the survivors.*

Dick Rodby, my great Hawaiian friend and Arizona Memorial benefactor.

you will not do that."

We left Tokyo on April 18th and traveled to Seoul and spent two days there. I am a nut about WWII history and we stood at the pass where North Korean soldiers fired their first shot at American soldiers in the Korean War in 1950. So on this trip I visited Pearl Harbor, Bataan in the Philippines and the spot where the Korean War began. In South Korea, we were driving in the country and went by a little grocery store that had Coca-Cola signs plastered all over the store. I thought, "Well, I guess we certainly are globalized."

Heading home, we stopped in Hawaii to unwind for a couple of days. I followed along with my camera and microphone as Secretary Butz visited sugar cane processors and the Dole Pineapple people. We were out in the fields as they were harvesting pineapple. It's an interesting, labor intensive process, with the fragile fruit picked by hand, then gently placed on a conveyor belt that takes them to a sorting area where they are carefully packed in bins and transported by forklift or truck to the packing sheds. During this trip, I met a gentleman who became a dear friend. Dick Rodby operated the Old Kemo'o Farm Restaurant just outside Schofield Barracks.

It had been in his family since 1920. The farm was created to raise food for the Barracks and it introduced commercial pig farming to Hawaii in 1909. In 1919, it added dairy farming.

After touring the pineapple fields, we went to Dick's restaurant. I was so impressed by the Kemo'o Macadamia Nut Cream pie, that's all I could think of when we planned our return to Hawaii a couple of years later. On that second trip, my wife and I drove to the restaurant where I told the waitress, "You know, I drove all the way from Chicago to have a piece of your Macadamia Nut Cream pie." A horrified look came over the waitress's face as she said, "Oh my golly, we don't have that on the menu today. We didn't make that pie today!"

Well, I guess she went back to the kitchen and told Dick about our long trip because, before we finished our meal, they had a Macadamia Nut Cream pie for us.

Dick was 11 years old on December 7, 1941 when he watched the Japanese Zeros fly over his house at 500 feet on the way to strafe and bomb Schofield Barracks. As a result, he became dedicated to making sure the world never forgot Pearl Harbor Day. In his later years, Dick spent a great deal of time volunteering to be part of the Arizona Memorial, welcoming visitors and answering questions. And particularly, he searched out people who he thought could have been there on that day in 1941. Dick found many interesting stories to share about the people he met. I've said many times that, until you had been to the Arizona Memorial with Dick Rodby, you had

At the Pearl Harbor Survivors' reunion in 2006 with Joan and Dick Rodby.

not really been there. He added a tremendous emotional feeling to the visit of the sunken battleship.

Our friendship with Dick and his wife, Joan, resulted in Gloria and I being invited as guests to the 65th commemoration of Pearl Harbor Day on December 7, 2006. I do find the history of war fascinating and since we had devoted some time and dollars to restoring the Arizona Memorial, we were on a VIP invitation list, which led to many additional opportunities that made it an unforgettable week.

We had the opportunity to spend time at several events with about 400 U.S. veterans who had been at Pearl Harbor the morning it was attacked. Also in attendance were about 20 Japanese Zero pilots who flew in the raid on Pearl Harbor. It was interesting to watch the people mix there. I would estimate that about 20% of our veterans would have nothing to do with the Japanese pilots.

I asked one American veteran how it felt to talk with the Japanese pilots. He said, "First of all, we were all very young people doing what our governments had ordered us to do. We were following orders. I'm 89 years old and I don't want to die with hate in my heart."

The presentations we heard during the commemoration included one from Joan Rodby, who was nine years old when the Japanese attacked and she talked about how life changed on December 7th. Beginning that day, residents had to cover their windows at night so no lights would be visible. She was in grade school and every student was given a gas mask and they had bomb drills where they would crouch under their desks, wearing their gas masks. Hawaiian officials were aware that a two-man Japanese submarine had been sunk just outside the harbor, and they were prepared for another attack.

Perhaps the greatest story, though, came on the night of the formal military ball. The ball was beautifully staged with the hosts wearing the military uniforms that had been worn in 1941. It was a lavish sit-down dinner and "PFC Gomer Pyle," Jim Nabors, sang the national anthem. That night, his television show rank was raised to Sergeant by a Marine General.

The story that stole the show began on December 6, 1941. It was the night of the Navy Ball at the Bloch Arena, and men and women from the Pacific Fleet were in their uniformed splendor. One of the couples had their 11-year-old daughter with them, Patsy Campbell. One of the most popular dances at the time was the jitterbug and there was a contest to find the best jitterbuggers. The orchestra leader knew that Patsy could really dance. He asked the crowd, "Isn't there a sailor out there who would like to jitterbug with this girl?" A 17-year-old sailor from Tennessee, Jack Evans, stepped up and said he wanted to. He and Patsy danced so well, they were crowned the Jitterbug Champions of the Navy Ball on that eve of the Pearl Harbor attacks.

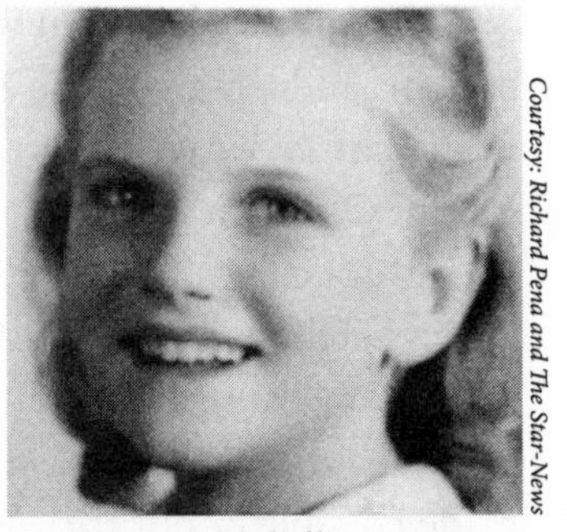

Courtesy: Richard Pena and The Star-News

Patsy Campbell in 1941.

Decades later, that "girl" wondered whatever happened to the sailor who danced with her. She began a search and eventually found a retired Navy veteran in Florida who said, yes, he remembered that young sailor and knew where to find him. When she contacted him, she was shocked to learn that he lived in Southern California, only 15 miles from her, and had for many years. That night at the 65th commemoration, they were both there. Their story was repeated for the crowd and they were introduced. The band started playing and the two former partners danced the jitterbug again, right down to the floor. They received a standing ovation that lasted four to five minutes. That was a thrilling moment, one of the great highlights of that event.

The 1941 Navy Ball Jitterbug Champions, reunited 65 years later.
Pat (Campbell) Thompson and Jack Evans.

Another highlight was the next morning when Tom Brokaw delivered his address about the "Greatest Generation." The day started early at Pearl Harbor; we were there by 5 a.m. The program started at 7:30 with the Marine Commandant taking charge as emcee. At 7:55 precisely, the moment of the 1941 attack, four Navy jets flew by at water level and then one of them went straight up in the Missing Man formation. As soon as that happened, a missile cruiser came steaming by with the entire crew in Navy white lined up at attention at the rail. It was a fascinating and emotional moment, all made possible by our friends, Dick and Joan Rodby.

By the way, if you've ever been to Hawaii and enjoyed a Happy Cake, you can thank Dick. In 1967, he put together some pineapple, macadamia nuts and coconut and invented the Happy Cake, which became a Hawaiian tradition, both for locals and tourists. His Kemo'o restaurant was famous for live Hawaiian music and was featured as a tavern in the 1953 Frank Sinatra film, *From Here to Eternity.* Dick passed away in January of 2012 just a few days shy of his 81st birthday. Joan still lives in Honolulu.

Our long trip ended in Chicago at midnight on April 25, 1974. We were exhausted but full of stories to share with the Midwest about history, foreign cultures and how farming is done halfway around the world.

RUSSIA

In 1983, Secretary of Agriculture John Block asked me to go with him to the U.S.S.R. for the signing of a trade agreement. My cameraman, Bob Varecha, and I flew commercially from Washington, DC to London, then boarded a U.S. Air Force DC-9 to fly from London to Moscow. In addition to our crew there were two uniformed Russian Air Force officers on board. During the first part of the flight, they stood at the front of the airplane staring at the seven of us in the Secretary's group. When we crossed the border into the Soviet Union one of the officers went into the cockpit to direct the U.S. pilots while the other officer continued to watch us to make sure Bob did no shooting out the window at the Soviet landscape below us.

When we arrived at Moscow International Airport, we were welcomed by the Ministers of Agriculture and Foreign Trade and were then driven in a military motorcade to the U.S. Ambassador's residence. I well remember that evening; Secretary Block and I enjoying a glass of wine while standing on a balcony looking at the Moscow skyline and saying to each other,

"What are these two Midwest farm boys doing in Moscow, representing the U.S. government?"

The Secretary set it up so my cameraman and I were part of his official delegation, which gave us access to places and people that we ordinarily wouldn't have had. But when we attempted to walk into the Kremlin to videotape the signing of a five-year grain agreement, security officials said, "Nyet." That changed when Secretary Block himself came back to the entrance and said, "These people are part of my official party and they must be with me to record the ceremony." The signing ceremony for a five-year grain agreement between the United States and the Soviets was held in the Kremlin. We were in a huge room, overflowing with all sorts of Soviet officials, including the Minister of Agriculture, Minister of Foreign Trade and Secretary Block and his delegation. After the signing, we were escorted into a reception room to celebrate the signings. It was a large, long room with a long table filled with many bottles of vodka and bowls of caviar. There was much toasting and it did get rather loud and boisterous before the event

Reporting from the Kremlin, 1983.

ended.

It is difficult to overstate how important that agreement was to American farmers. In the two years prior to the 1983 agreement, U.S. agricultural exports had fallen 20% and in 1983 amounted to $34 billion. Compare that to 2011, when our ag exports totalled $137.4 billion. With the signing of the 1983 agreement, America's producers truly moved into the global marketplace. Of course, our relationship with the Soviet Union has had its share of ups and downs. When Jimmy Carter was President, he embargoed wheat sales to the U.S.S.R. because he was going to teach the Soviets a lesson for the way they were treating Afghanistan. As we saw, the only lesson was that there were a lot of other wheat-growing countries that were more than happy to sell to the Soviets.

Being in Moscow was the one time in my career that I felt intimidated and was careful in choosing my words. When it came time to record the commentary I do on WGN Radio and on *This Week in AgriBusiness* called *Samuelson Sez*, I had decided on the theme: "America, thank God for what you've got because we live a free life compared to what's here." My cameraman had set up with the Kremlin looming in the background. Well, as we set up the gear, a crowd gathered and as I'm looking at the Russian people, I'm thinking, "Okay, how many KGB agents are in this crowd?" The more I looked, the more I re-thought the pro-American comments I was planning. My fear was that if I said something they didn't like, they'd take all the video we'd shot on the trip and we would go home empty handed. So, I caved and waited until I got back to the studio in the U.S. to say what I was going to say in front of the Kremlin.

CUBA

Fidel Castro loves to talk. He's famous for holding meetings with groups of foreign visitors where he just sits and talks for four or five hours and the foreigners are too polite to leave. Luckily, when we visited Cuba in October of 1999, we weren't another of his captive audiences. I met him and shook his hand and listened as a translator told me how he was glad we were there. I was part of an Illinois trade delegation headed by Governor George Ryan, who became the first governor to set foot in Cuba in nearly 40 years. The governor's intent was certainly to promote a trade relationship, but the U.S. government didn't like that, so it was re-characterized as a humanitarian mission. There were about 50 of us on the trip, including state lawmakers and officials, educators, religious and cultural leaders and an agricultural delegation that included leaders from John Deere and Archer, Daniels Midland. Uncle Bobby and his wife Christine joined us, as did my cameraman, Phil Reid.

The trade embargo imposed on Cuba in 1962 is ridiculous. It should have ended long ago. There it is, only 90 miles off our shore, more than 11,000,000 people ready to buy our rice, meat, milk and other commodities. Cuba produces sugar and some fruits, but they don't produce much in the way of meat or milk. We were driven to a dairy farm that was a commune farm. The manager almost had tears in his eyes when he said, "I wish I could bring the Holstein genetics from the U.S. This herd was started half a century ago with genetics from the United States, but now I can't improve them. Not only that, I can't import the grain to feed my animals to get top production out of them."

Talking with Cuban people, they said, "Oh, we love that Texas long-grain rice. That was our favorite rice. Now we're getting rice from Vietnam, and we don't like it. We'd love to get that Texas rice back into the market." Havana, one of the more progressive cities in our hemisphere before Castro took over, is one of the poorest with its citizens lacking many of the conveniences of everyday life that we in America take for granted.

To me, embargoes never work unless you are solely in control of whatever it is you're embargoing. But Cuba was able to trade with every other country in the world. Instead of buying from us, Cuba was getting its soybeans from Brazil, its wheat from Canada, its rice from Vietnam, its poultry from France. It was getting everything it needed, and here we are, ninety miles away, feeding the whole world — except for Cuba.

On the airplane going down, I sat next to Allen Andreas, who was CEO of Archer Daniels Midland at the time, and I asked, "How frustrating is it for you?"

"Orion, every day, we have ships coming out of the Gulf that go right by Cuba with everything those people need, and we can't stop."

Illinois agriculture, indeed all of Midwestern agriculture could benefit greatly from the removal of the embargo. Governor Ryan showed Fidel Castro on a map how easily grain, meat, tractors, bulldozers, trucks — pretty much anything — could be put on barges and sent down the Mississippi River into the Gulf. I have maintained hope through every administration since the embargo was imposed that it would be removed, but there are a lot of politics involved. There's heavy lobbying to maintain the embargo by the Cuban population in south Florida. George W. Bush wouldn't even bring the issue up because he needed the Florida vote so badly in the presidential election. But, after he won re-election, he did allow an adjustment in the embargo. If Cuba pays cash on the barrel head, we can sell them food. And we have sold a few billion dollars worth, but they'd be an even bigger customer if credit were allowed, and they'd certainly get that Texas long-grained rice that they really want in there. It just doesn't make sense. We imposed the embargo to get Castro out of there. He has outlasted nine presidents.

CHINA

In April of 1978, I went to China for the first time with an Illinois delegation appointed by Governor Thompson. This was before the U.S. had normalized relations with the People's Republic of China, but Governor Thompson had a feeling that President Carter was going to China shortly after that to open up trade, which he did in November of 1978. The governor said, "You know, there are about 800 million people in China and they're going to need food. They're going to need agricultural technology, they're going to need seed, they're going to need equipment, so we ought to go over there and let them know that Illinois has the goods that they're going to need." So he appointed 18 people representing the Farm Bureau, Farmers Union and some of the other commodity groups. And I was invited to go along. All I could take on that trip was an eight-millimeter camera because the Chinese had allowed an Italian film crew to come in the year before with a 16-millimeter camera, the broadcast standard at that time. They shot a documentary, took it back to Italy and put it together. It was devastating to China. So the thinking in China was, if anybody comes in with a sixteen-millimeter camera, look out for them, because they're going to hurt us. I found the best eight-millimeter camera I could find and we did come back with usable video of what it was like on Chinese farms.

We spent two weeks in China. Even though it was spring, there was still snow on the ground in Northern China. We had three guides and interpreters. One was a gentleman in his late forties who was a member of the Communist Party, Mr. Yuan, and two younger associates who were not members. It was not all that easy to become a member of the Communist

Recording an interview on the Great Wall with Warren Lebeck, president of the Chicago Board of Trade.

Party; you really had to work and prove yourself and the two younger men were trying hard to impress Mr. Yuan in hopes that they would be allowed to become members of the Communist Party. We got into some interesting discussions with Mr. Yuan, and we went back and forth at each other the whole trip about the benefits of capitalism versus communism — never angry, but each of us were making our point. He kept telling me that one of the benefits of communism was that everyone in China was equal. So we'd visit the communes — the farm communes — and you'd walk into one room with two shelves on the wall that were beds for the people who lived there, and you had a round hole in a raised concrete structure which is where they built their cooking fires and there was a table and a couple of chairs. Even at that point, they were beginning to curtail population growth to one child per family. In the Chinese culture, it was important to have a son, a male heir. Mr. Yuan wouldn't talk about it, but it was obvious that very often baby girls ended up in the irrigation canals because the farmers knew they could have one child and they wanted a boy.

It was an absolutely fascinating trip. I saw a world that I didn't realize existed. We stayed in the Peking Hotel where foreign guests stayed — that was the only place we were allowed to stay in Peking, which the capital was still called at that time. It didn't become known in the West as Beijing until the 1980s. The name never changed as far as the Chinese were concerned. To them, it was always Beijing; the westerners just spelled it differently for a long time.

The first day I'm there, I'm in my 14th floor hotel room and I opened the window and looked out at the street below me. It was filled with bicycles. As far as the eye could see, it was bicycles. And whoever designed the main thoroughfare didn't fully think it through. It was a ten-lane street and the four outer lanes on each side were for bicycle traffic. They put the vehicle traffic in the two middle lanes and it was a circus watching from fourteen stories up as these cars would come down the middle lane and then turn right in front of four lanes of bicycles. How they kept from killing bicyclists, I never did figure out.

Chairman Mao had died about a year and a half earlier. Of course, Christianity was deemed to be an unwelcome tool of Western troublemakers and during Mao's time, practicing Christianity publicly wasn't generally a path to a long life. Somehow, a Chinese minister survived the purge and led Easter Sunday services in Chinese. There were Africans, Germans, Norwegians and our group. There was also a Catholic church that served that community. So, in the morning, we worshipped our God, and then in the afternoon, we were driven to the tomb of Chairman Mao, where his body was on display. It was unbelievable to see Tiananmen Square filled with people, but all standing in military platoon-like files. They were civilians waiting patiently to see Chairman Mao. They were orderly. There was no moving around. They just stood in line and when it was their turn they'd be moved forward. Well, we had nine cars for our group. We had two Americans in the backseat of each car, and madmen behind the steering wheels.

We were driven right to the head of the lines. There were two lines going on each side of Chairman Mao's body. As I walked by one line, I looked across at the other line, which was all Chinese people, tears rolling down their cheeks. Some were sobbing. That night in my hotel room, I wrote a letter to my pastor saying, "Today I saw the Chinese worship their God, and I had the opportunity to worship mine."

As we made our way around China and I was seeing things that I'd never seen, the Chinese saw something they'd never seen: me. I'm already a tall person by American standards and I wore cowboy boots the whole trip which added a couple of inches. That made me a giant by Chinese standards.

On the Great Wall of China with an audience in awe of the giant American wearing cowboy boots.

Kids just stopped and stared because they'd never seen Americans let alone one as big as me.

One of the things that was kind of funny — at least it started out that way — involved Big Jim, a hog named in honor of Governor Thompson. Big Jim was a purebred Yorkshire boar that we loaded up in Illinois and took with us to China. Getting Big Jim into the country was quite an ordeal. We had to land at Tokyo for refueling before we went into Shanghai, and it was an interesting time.

China, at that time, had about 240 million pigs. Here in the U.S., 60 million is about the average number at any one time. But the Chinese pigs were just running loose. Running isn't accurate. The Chinese pigs were fat and lazy; all they did was eat and sleep. They were not being fed for food at that time; they were being used for fertilizer. Whenever and wherever a pig dropped its thing, somebody picked it up and put it on a field, if the pig wasn't already on the field. That Chinese pig, the Meishan breed, was the ugliest pig I'd ever seen. The best way I can describe it is maybe 200 pounds of lard hanging on a backbone with huge, droopy ears and wrinkled black skin. But they were prolific breeders with huge litters. In this country, the average litter size would be nine baby pigs. The Meishan had fifteen to twenty-five or more! The sows didn't have enough faucets to feed them all. That was astonishing to see. We had some animal experts with us, like Dr. Orville Bentley, Dean of the College of Agriculture at the University of Illinois as well as other university scientists, and they were astounded with the Meishan and its litter sizes.

We took our Illinois boar over there to give to an agricultural school with the idea that the Chinese would use it to breed with the Meishan sows, get a breeding program going, and then they'd begin raising pigs for meat and when trade relations were re-established, they would import corn and soybeans from us. That would be the beginning of our exports to China. So it was a goodwill gesture with a benefit built in for Illinois farmers. When we got Big Jim to the university's hog farm, which was a government hog farm — everything was government run — and we first laid eyes on the Meishan, someone in our group said, "We're going to have to put a paper bag over Big Jim's head before he'll make love to that." No paper bag was necessary because Big Jim sired a number of litters, but six months later, he was dead. The Chinese had a killer cat disease that was transmissible to pigs. We don't have it in this country, and so there was no immunity. John Block, who was on the trip as Director of the Illinois Department of Agriculture, has joked over the years that Big Jim died in service to his country. We did bring some Meishan pigs back to the University of Illinois for research, but the breed never caught on over here for a few reasons, including that its meat is so fatty. I've heard that it's considered a delicacy in Japan, but, my golly, that was one ugly breed of pig.

As you'd expect, our visit was carefully monitored by the Chinese Communist Party and it was pretty much dictated where we could go and what we could do. We went to a tractor factory in Changchun, which is a "little town" in far northeastern China of three and a half million people in 1978. In the 2010 census, the population was nearly eight million. The tallest building in town was three stories. At the tractor factory, I snapped a very unique picture — a picture of Hans Becherer, who later became CEO of Deere & Company at Moline, sitting on a red, Chinese tractor. I've kind of blackmailed him in a friendly way over the years because a green tractor guy doesn't sit on a red tractor.

While we were visiting the tractor factory in Changchun, we were staying in an old hotel they opened just for us. It was a mess, total disrepair, but as we walked around, we could see that it was a magnificent hotel at one time, but no more. There was no heat, and in northern China in April, it was cold... in the twenties. What we finally were able to determine was that the hotel served as a rest camp for Japanese officers when the army occupied that part of China. That's where the officers would go to spend their R&R time. At that time, they had the use of an indoor swimming pool. When we were there, it was filled with broken furniture. There was a huge rug, about 40 by 60 feet, in the entryway that you could tell had once been absolutely grand and beautiful, but it was so dirty and stained that there was no way it could have been restored.

The members of our group kept telling our guide, "We want to go to an agricultural university," and we got all sorts of reasons why we couldn't. "The roads are too bad." "It's too far." "We just can't take you." Well, we knew why. Their university system, at least the agricultural colleges, had been disrupted by the Cultural Revolution. The professors had been sent to the fields to work. We kept insisting, and finally they said, "We'll bring the professors to you."

Fourteen professors of agriculture arrived at the once-grand hotel where we were staying, and when they saw their American counterparts, it was like watching men in the desert dying of thirst suddenly finding water. All of them had been educated in the United States, and when we introduced them to Dean Orville Bentley from the University of Illinois, they literally grabbed him, and asked questions like: Is this professor still alive? Is this professor still at Purdue? Is this professor still at Minnesota? All of them had gone to school in the United States in the '40s, before the Communist takeover, and after returning to China had no contact with the outside world. They stayed until two o'clock in the morning. It was phenomenal to watch.

The next morning, two of them came back and brought soybean seed from their experimental farm. The soybean originated in China somewhere around the 11th century B.C. as far as scientists can determine. The seed showed up in the United States in the 1760s thanks to a sailor, but Henry Ford was the soybean's champion, spending millions on research and development that led to soybeans being used in the manufacture of Ford cars and for many other uses, including food. When the Chinese professors showed up at our hotel with their seed, that started a dialogue that continues to this day. Dean Bentley began a yearly trip to China where he worked with the Chinese on nutrition. After he retired, Dean Bob Easter picked up where he left off.

The farms that we were allowed to see were extremely primitive compared to anything in the United States at that time, and you know that they took us to their best farms. Really, it was impossible to make a comparison because there was nothing there from a structural standpoint, an infrastructure standpoint, a technology standpoint, that was like anything we had in the U.S. For example, the farm houses: Chinese farm families lived in one-room communal buildings. The buildings were about 200 feet long with single rooms side-by-side. Each room housed one family. The children went to school, but as they matured and if they passed tests showing they had real potential either as a student or athlete, they were taken away from their parents and sent to schools in cities so they could be properly educated. I managed to talk to one set of parents who spoke some English, and I asked, "Do your children go away?" And they said, "Yes." In some cases, they would never see their children again.

The method of production was all hand labor. The fields were very small. The farms were large, but they were actually groups of small farms that had been privately owned, but were taken over by the government. Each of the new, big farms was given an assignment. They had to produce so much rice or so much whatever the crop that they were assigned. The farmers had no input on what they would plant. That was all assigned from the central government in Peking, often thousands of miles away. Then, they had to meet that assignment, or production goal, and whatever they raised over it, they could keep to feed themselves and to feed the commune.

I mentioned Mr. Yuan earlier in this chapter, our Communist Party guide with whom I would playfully joust about the benefits of capitalism versus communism. As I looked at those one-room homes and the abysmal living conditions, I said to Mr. Yuan, "So, this is the way Chinese people live?"

"Yes," he said, and proudly added, "this is the way they live, and they're taken care of. They pay no rent and medicine is free." "What about Chairman Mao?" I asked. "Does he live like this?"

"Oh, yes, he certainly does."

He was starting to irritate me, so I kind of laid it on him pretty thick. I said, "Well, let me tell you how I live. I have a three-bedroom home. I have two automobiles. I use an airplane to get to many of my meetings."

He looked at me, and said, "Well then you are part of the four percent of the wealthy imperialists and the other 96 percent live in extreme poverty in your country. We know what your country does to people who are poor."

There was no convincing him otherwise, but I added, "I hope someday you can come to America, come to my house, and you can see not only me, but all of my neighbors, who pretty much live the same way. I am not considered a wealthy person in my country. I am considered comfortable, but not wealthy." He never did accept it.

During one of our trips within China, we flew on Chinese airliners. They were Russian airplanes, imitations of the Boeing 727. The first time we did, I walked on to the airplane, and as I got to my seat, I couldn't sit down. The seats were too close together, or I was too tall, whichever way you want to look at it. I could get halfway down in the seat, but that was it.

"Mr. Yuan," I said, "I can't sit."

He looked at my predicament and said, "Well, follow me." He led me to the front of the airplane, and there was a first-class section. I turned to Mr. Yuan, "I thought you said everybody was treated the same here. This is a first class section."

"No, no, no," he protested. "This is for military people. They can spread their maps out so they can work while they're flying."

I didn't believe him.

I do believe that our first trip in 1978 helped sow the seeds for what became a multi-billion-dollar-a-year agricultural trade between the U.S. and China. After we started trading with China, other countries in the region like South Korea, Japan and India, began importing huge amounts of our grain. Ironically, China, from which we got soybean seeds about 250 years ago, now buys more of our soybeans than any other country.

The Chinese feed it to their livestock and they feed it to humans, too. They take the soybean oil and use it for cooking and the diet. Milk had never really been part of the Chinese diet, but they've developed a taste for it, so you're seeing dairy farms go in, again using American know-how. China's corn acreage has expanded over the years, but the population is huge and the challenge to the Chinese government is immense. It can't move people off the farms because the cities are overcrowded. So, instead of mechanizing, they keep a lot of hand labor on those farms, and that limits production.

When we visited there in 1978, the thing they talked about more than anything else that they needed was fertilizer. And at the time, they were building 14 fertilizer plants. China has a 4,500-year history of just putting hog manure on their farm fields. Take a look at how their soil has been mined of all the nutrients it needs to grow corn, soybeans and other crops. If you're going to grow corn and soybeans, they take nitrogen, they take potash, they take phosphate — you've got to put it back. The Chinese reached the point where they were using more fertilizer, but they still hadn't mechanized to the extent necessary to make their farming efficient.

Since that first trip in 1978, I have been back to China ten times. It's a different country every time I go back. The biggest changes are in the cities — Shanghai, for example. When I was there in '96 with Secretary Glickman on his trade mission, the mayor of Shanghai took us on a boat tour of the harbor. She said 20% of all of the building cranes in the world were at work in Shanghai. I didn't doubt her because we stayed in a new hotel that was about 50 stories tall, and looking out my window, that's all I saw — cranes. I saw all the little homes that were being torn down, and the cranes were putting up shiny, new skyscrapers. There were a lot of changes in the cities, much more than what we've seen on the farms. They are still collectives. You still don't see much mechanization and it isn't because the machinery isn't available. John Deere has a manufacturing plant in China, but the farm fields are so full of ditches, either to irrigate or to take water away for drainage, and those ditches are big enough that you just don't roll combines or tractors over them. They would have to restructure the farmland in many instances to bring in the kind of equipment that is commonplace on an American farm. If they do that too quickly, then it's back to the question raised earlier: what does the government do with all the laborers? Sending them to the cities isn't an option. It has to be a tightrope for the Chinese government. It really does. Of course, we hope they succeed because it's a huge and important market for us. And American agriculture has heavily invested in China. Not only Deere, but ADM has built processing plants there.

Another by-product of the 1978 trip and subsequent agricultural trips to China and by the Chinese to the United States is that the Chinese have developed a taste for meat. And, again, Chinese farmers can't always raise enough meat to feed everyone, so it's an opportunity for U.S. producers. Pork exports were in the $4 to $5 billion range from 2008 through 2011, but varied greatly as the price of Chinese pork rose and fell and because of the H1N1 ban on U.S. pork imports.

The Chinese don't eat as much beef as they do pork and poultry. During that first trip over there, I was a guest of honor at one of the meals, and was served chicken. The whole chicken, head and feet included. I ate the head, but passed on the feet, but feet, or paws, are a regular menu item for the Chinese, a delicacy. Leave it to an American capitalist to seize on an opportunity! On a later trip, I was sitting at an agricultural event and not far away was a gentleman from Washington State who was making a huge living selling chicken paws to the Chinese. We throw them away and he was shipping thousands of tons to China and making a huge living doing that. Growing up on the farm, the joke was that Mom would cook everything but the pig's squeal, the cow's moo and the hen's cluck. But, there were some things where I drew the line and chicken feet were one of them. I knew where they'd been.

THE U.K.

One of my favorite countries is the United Kingdom. During one trip in 1979, we did the WGN *Noon Show* live at 6 p.m. from the Royal Agriculture Society of England show in Stoneleigh, England. Someone from the royal family would always show up at these events and at that particular show, Princess Margaret was there and made a comment I'll never forget.

Big O with Big Ben in the background.

I spent 30 minutes at a reception talking with her. Her husband was into horses and equestrian activity. I asked, "Your Highness, does your husband go with you when you go on your royal tours or does he stay home on the farm?"

Her response: "No, he stays home and takes care of the horses while I do my royal stuff." Hah! "Royal stuff." Funny line from the stiff upper lip of a British royal.

Also on that busy trip, as guests of the British government, we were given tours of the Houses of Parliament, the British Agricultural Export Council, the Agricultural Engineers' Council, the *British Farmer and Stockbreeder* magazine, the Banbury stockyards and various food distributors. We were shown an embryo transplant demonstration and also went to Scotland, where we toured the North of Scotland College of Agriculture in Aberdeen.

In 2007, Gloria and I flew to London for a quick trip and it was at a time when the dollar was at one of its lowest points versus the British pound. When I went to the currency exchange window to trade dollars for pounds, I gave the teller US$200. She gave me back about 87 BP! That meant when we checked the price of an item in a store or the price of a meal on a restaurant menu, we had to more than double the pound price to realize what it was costing us in American dollars. Breakfast at one hotel was 24 BP, which in U.S. dollars was more than $55.

Of course, while it was a financially challenging time for American tourists in Europe, it was a good time for American exporters. A weak dollar definitely gives us a competitive advantage in the world market, particularly in agricultural products like grains and soybeans.

Another thing I noticed during my trips to the United Kingdom was that government officials are not held in very high esteem, regardless of the country. It makes me wonder why anyone wants to run for public office. In 2007, the relatively new leader of the British government, Prime Minister Gordon Brown who succeeded Tony Blair, was in the headlines daily because of a scandal in campaign contributions and his inability to deal with that kind of a scandal. Sound familiar?

On the agricultural front, foot-and-mouth disease was back in the headlines because of the second leak of virus in three months from a laboratory that is licensed by the government to manufacture an anti-virus to control the disease. These two leaks did flow into streams that ultimately led

Visiting with the U.K.'s Minister of Agriculture, the Right Honourable Peter Walker in July, 1979.

ABOVE: *Longtime television production partner Phil Reid and I are discussing our strategy at the Royal Agriculture Show in England.*
BELOW: *Interviewing the Earl of Mansfield as his prized Highland cattle look on at the Balboughty Farm in Perthshire, England. July, 1979.*

to an outbreak and once again, the culling of some cattle herds in a country that was devastated by foot-and-mouth disease a decade ago. Farmers were extremely angry and blamed the government for a lack of oversight and control of the laboratory that once again threatened the future of the livestock industry in the U.K.

Also during that 2007 trip, a columnist in a London newspaper flatly stated in alarming terms that the planet will soon run out of food and we are on the brink of world-wide starvation. His reason? We are converting far too much grain intended for human food and livestock feed into energy for our gas guzzling automobiles. Of course, we've heard that argument on this side of the Atlantic, too.

So, with all these challenges, did we have a good time? No, we had a great time! I love England for its sense and appreciation of history and of the 43 countries I have visited, it ranks second on my "favorite" list after Norway, the country of my ancestry. For me, spending time and money on travel is one of the best investments I can make.

ARIZONA

My first visit to Arizona was in 1979. I was invited there for three weeks by International Harvester for the introduction of its radical, new tractor, the "2 Plus 2." It was an articulated tractor that turned in the middle and was a major change from what we were accustomed to in tractors. IH made a huge production out of it, taking over a farm in the Valley of the Sun and renting the Dick Van Dyke Studios. We would give presentations to dealers each morning at the studios and in the afternoons they would go out to the farm and drive the tractor. IH brought in 5,000 dealers over a three week period, so every other day, we would get a new group of 500 dealers. I was the emcee of the morning program. We had singers and dancers from Broadway and the entire presentation was done to the theme of *A Chorus Line*, re-written to introduce the new tractor. It was a spectacular event and, again, my first visit to Arizona. It's when I fell in love with the desert.

The first night of the event, the 500 dealers were taken to Rawhide, an old Western town, a tourist attraction, where they were entertained. Danny Davis and the Nashville Brass performed, as did Troy Nabors, a comedy trick roper who had a trained mule. That mule could do anything and everything that a mule would probably never do in its natural life. Well, I got a chance to

Good Arizona friends Troy and Jan Nabors.

talk with Troy and learned he was a retired rodeo clown who grew up poor in Oklahoma. He decided to get into the rodeo world and did it well; he's a member of the Pro Rodeo Hall of Fame. He became good friends with Rex Allen, Roy Rogers and many other Western cowboy stars. Over the three weeks, we got to know each other pretty well, but when the event was over, I went back to Chicago and forgot about Troy.

Five years later, when I decided to build a small home in Arizona, I wondered what had happened to Troy. I looked him up in the phone book, called him, and it was as if we had just seen each other the day before, not five years earlier. The friendship was renewed and over the years, we've spent a great deal of time with Troy and his wife Jan.

Troy is one of the best storytellers I've ever known. He appeared in several movies, the best known of them was *Raising Arizona*. Troy and Jan attend every party we host in Arizona and he always enthralls the guests with his stories of life on the rodeo clown circuit and the characters he's known over the years.

It's interesting how friendships develop. Troy and Jan are certainly in that unusual category of people I probably would not have known had I not had the opportunities I've had in this job.

ALASKA

In 1990, I was booked to go on an Alaskan cruise. I was anxious to go because I had heard there was farming in Alaska and was curious to see for myself how it was done in the far, far north and what was raised.

My research indicated that one farming area was in the Matanuska-Susitna Valley, which includes the towns of Palmer and Wasilla. Legend had it that cabbages grew so big they had to be harvested with chainsaws! Well, of course I wanted to see for myself, and wanted to bring back video proof. So I left a few days earlier than scheduled for the cruise and took a cameraman with me to hunt for these giant cabbages as well as look for other agricultural stories.

In researching the Matanuska-Susitna Valley, known in Alaska as the Mat-Su Valley, I was surprised to learn that some of my "people" had pioneered modern farming there. Norwegians and other Scandinavian farmers from Wisconsin, Minnesota and the Dakotas were lured to the territory (not yet a state) of Alaska beginning in 1935 with the promise of rich, loamy farmland. The Alaska Rural Rehabilitation Corporation created the Matanuska Valley Colonization Project and offered transportation to Alaska, a house, 40 acres of unimproved land (trees had to be cleared) and necessary tools and equipment. About 200 Midwestern farming families accepted the offer, which included long-term, low-interest loans from the federal government. In the years since, they cleared over 6,000 acres of land, built dairy farms, started hog, poultry, sheep and beef operations and learned how to make hay (and giant vegetables) when the sun shines, which it does 18 hours a day during the summers. The Mat-Su lies on about the same latitude as Oslo, Norway. Temperatures in the summer average near 60 degrees. Temperatures in the winter average about 13 degrees. The typical growing season is only about 110 days. Crops include potatoes, hay and barley, and some unbelievably big vegetables. Several world records belong to farmers from Alaska, including a 19-pound carrot, a 39-pound turnip, a 63-pound bunch of celery, a 65-pound cantaloupe, a 76-pound rutabaga, a

Mike Dunham, Anchorage Daily News

"The Beast," the world record cabbage, grown by an Alaskan dentist and harvested with a chainsaw.

97-pound kohlrabi, a 106-pound kale plant and a 127-pound cabbage!

Two and three generations later, those hardy Scandinavians are still farming the area.

My visit was at the height of the Northern Spotted Owl controversy. Environmentalists claimed the owl was in danger of extinction because of logging operations across the Northwest and in Alaska. I covered the controversial story extensively and I learned of an organization in Ketchikan, Alaska formed by lumbermen's wives to respond to the emotional negative claims and protect their jobs and the industry. Ketchikan was the first stop on our cruise and we made arrangements to do an interview with the ladies. My cameraman, Bob Varecha and I were the first people off the ship and on the dock waiting to be met by the ladies. While we waited, a young man came walking toward us. He looked like he was straight out of central casting for lumberjacks. This big, strapping young man had apparently seen other TV crews land on the dock and was curious about our intentions.

"Where are ya coming from?" he asked as he scowled at our camera gear.

"Chicago. We're here to do a story on the spotted owl/timber controversy."

"Which side are you on? Are ya for the timber industry or are ya for the spotted owl?"

I was pretty sure which side he was on. It wouldn't have surprised me if he had a bumper sticker on his truck reading, "I Like Spotted Owl — Fried."

"Well," I said, "we're from a show called *U.S. Farm Report* and we're pro-timber."

"I'm glad you said that," he said, "because otherwise, I was going to have to go shoot a spotted owl and come back here and shove it up your a**!"

Sometime later back in Chicago, I was on the air during the WGN *Noon Show* and for some reason the spotted owl controversy came up. I asked, "Why is it that humans need very little space to procreate — they can do it in a 12 by 12 room or even in the back seat of a car — but the Northern Spotted Owl needs 25 million acres? Well, the phones lit up and after taking a few profane calls from angry owl lovers, my assistant, Lottie Kearns, refused to answer the phone.

Reporting from the Royal Agriculture Society's Royal Show at Stoneleigh, in Warwickshire, England.

In 1999, I accompanied a United States Grain Council mission to Southeast Asia. In the photo above, the three men in the right half of the photo are Lyle Cook, former chairman of Iowa Corn, Ken Hobbie, former USGC CEO, and Cary Sifferath, the USGC Regional Director for the Middle East, Africa, and Europe. They are celebrating the signing of a memorandum of understanding with Suba Indah, at the time, a new corn wet milling company in Indonesia. In return for the Indonesian company purchasing a certain amount of U.S. corn, the USGC agreed to train Suba Indah staff on wet milling of corn for corn starch production.

Planting a tree on a Jamaican tree farm.

Ole and Lena got married in their little country church. As they headed down the highway for their honeymoon trip, they were nearing Milwaukee. Ole put his hand on Lena's knee.

Giggling, Lena said, "Ole... you can go farther if ya vant to."

So Ole drove to Green Bay.

To Orion Samuelson
With best wishes, Ronald Reagan

Chapter Eight

Presidents

On a foggy, drizzly spring night, I stepped onto the front portico of the White House and thought about my father because it was May 7th, 1972, his 73rd birthday. I said to myself, "Dad, I bet you never imagined your son would be having dinner in the White House, but here I am." Of course, given Dad's opinion of the people in Washington, he might not have been all that impressed.

But I was. I had goosebumps when I met my first President of the United States. I am proud to be an American and the specialness of those meetings that my job has provided was never lost on me. I may not have gotten goose bumps with every president I met, but I certainly felt honored and I appreciated each opportunity.

DWIGHT D. EISENHOWER

"Ike" was the first one. I was working in Green Bay at WBAY where I had joined the National Association of Farm Broadcasters. Each year, we convene in Washington, DC where we meet with members of Congress from farm states, meet with the Secretary of Agriculture and his staff and visit the White House. President Eisenhower had an Angus herd on his farm near Gettysburg, Pennsylvania and liked to talk agriculture when he met with the farm broadcasters. He shook my hand — definitely a goose bump moment — shaking the hand of one of the great war heroes in history during my first trip to the White House.

JOHN F. KENNEDY

My first day on the job at WGN was the day of the Nixon-Kennedy debate at CBS in downtown Chicago. That was the televised debate in which makeup, or the lack of it, did Mr. Nixon in. Shortly before that, while I was still working in Green Bay, Senator Kennedy came to town to campaign and held a news conference, which I attended for WBAY. I asked a couple of agriculture-related questions. After the news conference, a man in a suit came up to me, one of JFK's aides, and asked, "Do you have a few minutes?" I said, "Yes, why?" "The senator would like to sit and talk to you about agriculture because he is not that well versed in what the issues are here in the Midwest." So, I found myself sitting in the bar at the Northland Hotel in Green Bay having a drink with Senator Kennedy and talking agriculture. Actually, he talked very little, asking many questions and listening. Given his wealthy roots in Boston, he probably knew less about agriculture than most presidents. But, it turned out that he was a great listener and a quick learner.

One of JFK's aides sat with him and took notes. I think I had a beer and I forget what the senator had. He was a delightful person with a very warm personality and we talked for about 25 minutes. Mr. Kennedy wanted to know about agriculture because he knew the Midwest was going to be critical to his candidacy and he played to it on the campaign trail, declaring, "The family farm should remain the backbone of American agriculture... the decline in agricultural income is the number one domestic problem in the United States." President Kennedy extended a program started by President Eisenhower, renaming it "Food for Peace," using food produced by American farmers to feed people around the world whose friendship we desired. He brought back the Food Stamp program and attempted to get legislation passed that would have helped raise farm prices, but because of his assassination, we never got to find out whether his plan would have worked.

LYNDON B. JOHNSON

Much of what JFK started, LBJ tried to finish. President Johnson grew up on a farm in Texas and later owned a ranch near Austin. He was known for his forceful skills of persuasion and when he and Secretary of Agriculture Orville Freeman were pushing through the five-year Food and Agriculture Act of 1965, Johnson didn't pull any punches, warning members of Congress that if they didn't pass his farm bill, 20% of farmers would go bankrupt and the American economy would suffer a crippling blow. He said, "We know from bitter experience that depressions are farm-led and farm-fed and we are not going to repeat that experience." At the White House bill signing, which I attended, LBJ praised the "miracle" of American agriculture.

RICHARD M. NIXON

In 1972, President Nixon invited me to have dinner with him at the White House. He invited 159 other people, too. Norman Borlaug, one of my personal heroes mentioned elsewhere in this book, was being honored. He was a Norwegian farm boy from Iowa who won the Nobel Peace Prize, so President Nixon decided there should be a day to salute Mr. Borlaug for his accomplishment and, while Mr. Nixon was at it, he made the event a salute to American agriculture. It was quite a sight: tractors and combines parked on the White House lawn.

That evening, the President hosted a formal dinner in the White House. My wife was with me, but we weren't allowed to sit together. They put spouses at different tables. At my table was Bob Dole, the senator from Kansas who became a good friend and Glen Campbell, a huge star at the time who was the entertainment for the evening. It was an interesting night, with the 21 violins of the Air Force Strings coming into the dining room and entertaining us, and then we went back into the reception room which had been set up theater style and it was show time. And who was the Master of Ceremonies? Richard Nixon, who was not a natural at it. He introduced Glen Campbell and after his set, the Marine Band played for dancing until midnight.

During my dinner conversation with Glen Campbell, I learned that he, too, had been a farm boy and that we were both staying at the Hay-Adams Hotel, just a block away from the White House. Glen said the President and Mrs. Nixon had invited him and his wife to join them after the show in their private quarters, but after that, they would be back at the hotel. Glen suggested I call his room later that evening, which I did, and he invited me to join him in his suite. We spent another two hours sharing stories about growing up on our farms.

RONALD REAGAN

We met several times, including when I emceed the advance program for his 73rd birthday party in 1984 at Dixon, Illinois, his boyhood home. He was born down the road about 25 miles in Tampico, but was raised in Dixon. The party gave me the opportunity to spend time backstage with the President where we chatted mostly about the radio business. Before he became a famous Hollywood actor, he worked in radio, starting at WHO Radio in Des Moines, Iowa. During my conversation with him in Dixon, the President of the United States told me he really enjoyed his time on the radio and still missed it.

Nobody could command a room like Ronald Reagan. He was called the Great Communicator for good reason. In 2011 it was my sincere honor to induct Ronald Reagan, posthumously, into the National Broadcasters Radio Hall of Fame.

GEORGE H. W. BUSH

"Bush One" came to a couple of the Farm Progress Shows where I introduced him to the tens of thousands of farmers, ranchers and other agriculture people in attendance. Later, I did a one-on-one interview with him in the White House.

My most cherished memory of President Bush is from an encounter I had with him at a hotel in Beijing, China. For years I served as the emcee for the World Food Production Conference, hosted by IMC Fertilizer, based in Northbrook, Illinois. The conference moved to a different city each year and featured outstanding speakers. At the Beijing conference, former President Bush was the keynote speaker.

The chairman of IMC hosted a private dinner with the President and Mrs. Bush. I was included in the 15 guests. Before dinner, Mr. Bush and I were among a group of men standing with our cocktails and carrying on a conversation when the President looked at me and said, "Orion, you have to help me. There was a well-known farm broadcaster who worked in the Department of Agriculture when I was in the White House, and for the life of me, I can't remember his name." I quickly said, "It was Herb Plambeck who

is now in his late 80s and still doing some radio at WHO in Des Moines." President Bush said, "I am glad to hear he is still active. Let's send him a postcard from Beijing." He turned to the IMC officials and said, "Excuse me gentlemen, but Orion and I have to do something."

We walked over to a desk in the corner of the room; he pulled a postcard out of the desk drawer and wrote, "Hello, Herb; Orion and I were talking about you in Beijing and want to send our best wishes." He signed his name to it and then turned to me, handed me the pen and said, "You sign it." I said, "Oh, no, I can't put my signature on there." He said, "Yes, you will. And it is up to you to see he receives the card."

That told me a great deal about the human side of George H.W. Bush. Herb Plambeck cherished that card until he died.

On my 75th birthday, the President sent me a letter, which I have included in the "Letters" chapter of this book.

BILL CLINTON

I never met Bill Clinton when he was president, but I sat in on several agriculture listening sessions that he'd hosted during his campaign, including one in Iowa, which was a day-long event. I was impressed that he sat there for about six hours, listening, taking notes and asking questions. I was impressed with his ability to listen and participate, because you'd think that he might just walk in, say hello, and then turn it over to someone else and leave, but he didn't, he was fully involved.

GEORGE W. BUSH

We met in Peoria during his 2000 campaign for president. As I was interviewing him, I asked how important the farm vote was to him. "Well," he said, "I'm talking to you, that should tell you." After the interview as he was walking away, I said to him, "Governor Bush, I just want to tell you that when your dad was President he did a one-on-one interview with me in the White House." He wheeled around and said, "I'll do the same for you." He never did, because he had hardly been in office when the 9-11 attacks happened and there were too many things going on. The only time I was in the Oval Office during his presidency was when he honored the four Outstanding Young Farmers of that particular year. That's a program I've overseen for over 45 years. We broke protocol, because for an event like that, the White House would never have allowed a news reporter in the office while the President talked to the farmers. But I guess he gave me dispensation when

he heard that it was me and remembered that we had done an interview with the promise of a one-on-one. Each president is allowed to decorate the Oval Office as he sees fit and Mr. Bush gave us a detailed tour, shown on the previous page, explaining the statues, the rug, the furniture and everything else that he had selected for the eight years he occupied the office.

BARACK OBAMA

I have yet to meet President Obama, but I did meet him when he was a United States Senator from Illinois, the job I briefly coveted, in 2004.

HARRY TRUMAN

I was just graduating from high school when President Truman was leaving office, so he had long been retired when I met him. Of course, he wasn't supposed to be president, according to the infamous 1948 headline in the *Chicago Tribune* declaring Dewey's victory. But the Trib was wrong and Truman was a friend of agriculture during his seven years in office, perhaps because he spent most of his younger years living on a Missouri

With President Harry Truman at his home in Independence, Missouri. 1965.

farm. In 1965, when I was president of the National Association of Farm Broadcasters, our summer convention was held in Kansas City. I drove to President Truman's home in Independence, Missouri to present him with an Honorary Membership Award.

President Truman was known for his early morning brisk walks in Washington, DC, with Secret Service men struggling to keep up with him. It was a bit sad to see the President slowly shuffle out of his office for the presentation and the photo. He was most gracious and articulate in his thanks.

Ole and Lena went to the county fair where there was a pilot selling rides in his open cockpit biplane. Ole asked the pilot, "How much for da plane ride?"

"Ten dollars for five minutes," said the pilot.

"Oh, dat's vay too much," said Ole and he began to walk away. The pilot called after him, "I'll make you a deal. If you and your wife can ride for five minutes without making a sound, then the ride will be free. But if you make a sound, you'll have to pay $10."

Ole and Lena decided it was a good deal and they climbed into the plane. The pilot did his best to scare Ole and Lena with loop-de-loops and flying upside down, but there was not a peep from either of them. After they landed, the pilot said to Ole, "I have to congratulate you for not making a sound. You are a brave man."

"Oh, yah, maybe so," said Ole, "but I gotta tell ya, I yust about screamed when Lena fell out."

Sharing a laugh with Secretary of Agriculture Earl Butz.

Chapter Nine

The Secretaries

One of the biggest, most important jobs in Washington, DC is that of Secretary of Agriculture. The USDA is the sixth largest employer in the federal government with approximately 110,000 employees. They work in more than 25,000 buildings around the world. If USDA were a corporation, it would be the sixth largest in the country. It writes enough loans that it would be the seventh largest U.S. bank. The secretary oversees one of the most diverse and challenging missions across all of government. USDA has responsibility for farming, rural development, research, protection and conservation programs; international food and agriculture trade; the nation's nutrition programs including food stamps, school lunches and breakfasts, and the Women Infants and Children (WIC) program. The USDA also contains the U.S. Forest Service.

When Tom Vilsack, the former governor of Iowa, was sworn in as Secretary of Agriculture in January 2009 for the Obama administration, he became our 30th secretary. As I looked over the lineup of his predecessors, I realized that during my role as an agricultural broadcaster I have known, interviewed and worked with 16 of the 30 secretaries. Some, by virtue of the short amount of time they served, were mostly placeholders. But most made significant, ground-breaking contributions to American agriculture.

My first Secretary of Agriculture was Ezra Taft Benson. When I began working at WBAY-TV and Radio in Green Bay in the 1950s, I joined the National Association of Farm Broadcasters. Our summer meeting was always held in Washington, DC and that's where I met Benson in 1958. He was an Idaho farmer and county extension agent who had been appointed Secretary of Agriculture by President Dwight D. Eisenhower on January 21, 1953. He served the full two terms of the Eisenhower administration. Benson opposed government price supports and aid to farmers, arguing that it amounted to unacceptable socialism, but with farm commodity surpluses increasing and farm income dropping, something needed to be done. Benson's USDA designed a voluntary land retirement program which took acreage out of production, thus lowering the supply and raising prices. The Soil Bank helped maintain farm income and also conserved soil. While the Kennedy administration decided to go a different direction in 1961, a key aspect of the Soil Bank survives today: the Conservation Reserve Program, where environmentally sensitive farmland is converted back to its natural state or is planted with long-term cover crops to control soil erosion, improve water and air quality and develop wildlife habitat. After his eight years as Secretary of Agriculture, Benson became a leader in the Mormon Church and from 1985 until his death in 1994, was the church president.

Telling Secretary Freeman what Midwest farmers were thinking.

Orville Freeman was next. He was Secretary of Agriculture in the Kennedy and Johnson administrations and is probably best remembered for initiating the Food Stamp program. Freeman was the last secretary to serve a full eight years, although he did so under two presidents. Appointed in January of 1961 by JFK, he served until Johnson left office in 1969. Previously he was a three-term governor of Minnesota. When I did television interviews with Freeman, we always were asked to shoot him from the right side because as a Marine during World War II, he was wounded by Japanese bullets, losing part of his jaw on the left side of his face.

Following him in January of 1969 was Clifford Hardin, who was raised on a farm near Knightstown, Indiana and attended Purdue on a 4-H scholarship. When President Richard Nixon appointed him in 1969, he had been chancellor at the University of Nebraska since 1954 and Cornhusker fans should thank him for the football powerhouse the university became. Hardin hired Bob Devaney, the legendary coach. I would best describe

The media-shy Clifford Hardin being interviewed in 1970.

Clifford Hardin as soft-spoken and very uncomfortable with the media. It was obvious that he did not enjoy news conferences and the one-on-one interviews were painful experiences. He's perhaps most remembered for putting together a deal that limited federal subsidies to any one farm to $55,000 on each of three basic crops: cotton, wheat and feed grains. The 1970 farm bill also included his "set-aside" plan which called for farmers to agree to leave a certain percentage of their land idle to qualify for federal payments and price supports. The major farm groups, which were accustomed to having great influence on agricultural legislation, opposed the bill, but it passed with bipartisan support in both houses. A long slump in farmers' incomes continued during Mr. Hardin's tenure, and he was replaced by one of my all-time favorites, Earl Butz in November 1971.

Earl L. Butz was another Indiana farm boy and Purdue grad. Just the opposite of Hardin, he was a great communicator who loved interacting with the media. He had a terrific ability to communicate the story of agriculture to the non-farm public. He was famous for bringing a loaf of bread with him to his media appearances and when he appeared with me several times on my morning show on WGN-TV, he'd pull out two slices of bread from his loaf and say, "Orion, this is what the American farmer gets from this loaf of bread. The rest of the money goes to everybody who handles the wheat from the time it leaves the farm until it gets to your breakfast table as toast." Earl

Secretary Earl Butz and his loaf of bread.

Butz revolutionized federal agricultural policy and re-engineered many of the farm support programs. He abolished a program that paid corn farmers to not plant all their land. His mantra to farmers was "get big or get out," and he urged them to plant corn and other commodity crops "from fencerow to fencerow."

One of the most memorable phone calls I received was from Secretary Butz. I was at the Illinois State Fair in August of 1976. At 8:00 a.m. the phone rang in my hotel room and it was Earl Butz telling me that Richard Nixon had just resigned the presidency, Gerald Ford was being sworn in as president and that he, Butz, would continue to serve as Secretary of Agriculture. He was a great storyteller, but it did get him into trouble. Because of a story that I felt was totally misinterpreted he was forced to resign on October 4, 1976.

The Secretary of Agriculture with the shortest tenure was John Knebel. With Earl Butz's resignation, Knebel was appointed secretary on November 4, 1976 and served the remaining two months of Gerald Ford's term.

When President Ford lost the election and Jimmy Carter moved into the White House, Bob Bergland was appointed secretary. He was a Minnesota farmer and congressman who served all four of the Carter years and left in 1981. Bergland went through some of the most tumultuous years a Secretary of Agriculture had experienced. In the winter of 1979, about 5,000 farmers from as far west as Colorado converged on Washington, DC. Some 1,500 of them drove their tractors to the capitol in a snowstorm to protest American farm policy and tell their tale of financial problems to the media and the DC policy-makers. Their long, slow convoy across the country gained support along the way and by the time they rolled into Washington, they had quite a head of steam. But Bergland was critical of the demonstrating farmers, suggesting that some were greedy and others were just looking for publicity for their new organization, the American Agriculture Movement. They blocked DC traffic with their tractors and some stayed for weeks, demonstrating and lobbying. They were angry at Secretary Bergland's comments and it was a challenging time for Bob. He's retired on his Minnesota family farm and we continue to be friends, sharing Scandinavian stories.

When Ronald Reagan was elected president, the list of potential secretaries of agriculture included my name. I was pretty far down the list, and as I've mentioned elsewhere in this book, I took the advice of Earl Butz and others and stayed behind the microphone rather than mount any kind of campaign to get appointed. It came down to the wire between Richard Lyng from California and John Block, a hog farmer from Galesburg, Illinois.

With Secretary of Agriculture John Block in his Washington office.

That race produced an interesting telephone call to my office one afternoon when I picked up the phone and the voice on the line said, "Hello, Orion. This is Bob." It took me a couple moments to realize it was Senator Bob Dole calling. He continued, "I've been asked by Mr. Reagan to offer my advice on his choice for Secretary of Agriculture and while I know about Dick Lyng because of his previous service at USDA, I don't know much about this Block fellow from Illinois. What can you tell me?" We had a 15-minute conversation and I concluded by saying, "Both men would make an excellent Secretary, but Block is an active farmer and with the farmers' anger against USDA at a very high level right now, he might be a better choice. "Jack" was selected and served from January 1981 to February 1986, right in the middle of some very difficult financial times for American farmers. Under Jack's leadership, the Russian grain embargo imposed by President Carter was removed and a greater emphasis was placed on developing export markets.

Bush appointee Clayton Yeutter.

In 1983, with record high levels of grain stocks and very low prices, farmers were in a bind. Jack helped design and implement the PIK (Payment In Kind) program under which producers who agreed not to plant portions of their land were paid with government surplus commodities. It took nearly 50 million acres out of production, and combined with a drought that hit the Midwest corn and soybean belt in the late summer of 1983, resulted in much lower stocks and higher prices, which helped ease the economic stress being felt in rural America. There are many stories to share about Jack Block. I knew him first as a farmer, then as Director of Agriculture for Illinois, then as USDA Secretary and today as a good friend. We share dinner whenever I go to Washington. When Jack left the office, Dick Lyng was appointed by Reagan and served out the rest of the term.

Then, when George H. W. Bush became President, he appointed Clayton Yeutter as Secretary of Agriculture on February 16, 1989. Clayton

On the air in WGN's Studio A with Secretary Dan Glickman.

was a Nebraska farm boy who got his law degree from the University of Wisconsin-Madison. I had known Clayton for many years because he had served in a variety of assistant secretary positions within the USDA in the 1970s and from 1978 through 1985, was president of the Chicago Mercantile Exchange. He left the CME to become U.S. Trade representative during the last four years of the Reagan administration. During that time, I was invited to accompany him on a trip to Japan for a meeting with trade officials to convince them to open their market to American beef. Even though consumers in Japan wanted U.S. beef, the government imposed strict restrictions on our beef imports because they wanted to protect the small number of beef producers in Japan. I vividly remember sitting in this large conference room with a very long table, the Japanese Minister and his many aides on one side of the table, Mr. Yeutter and his aides on the opposite side. All the conversation was done through interpreters and I noted many smiles and head-bowing. At the end of the meeting, I said to Clayton, "It looked like a successful meeting with all the smiling on both sides of the table." Clayton said, "What you probably didn't see is that I was smiling with gritted teeth!" Yet, that meeting led to negotiations that ultimately did open the market to U.S. cattlemen and Japan became our No. 1 beef export market for many years. Credit Clayton for getting that done. After leaving the secretary's job on March 1, 1991, he became chairman of the Republican National Committee for a year, then was the White House Domestic Affairs Advisor before he left public service and went to work as a lawyer and lobbyist. Clayton is a very good friend to this day.

For the rest of the Bush presidency, Edward Madigan, a ten-term congressman from Lincoln, Illinois served as ag secretary. Among Ed's distinctions was that he was the first Roman Catholic to be appointed Secretary of Agriculture. I had gotten to know Ed while he served on the House Agriculture Committee from 1983-1991. He was in the secretary's job for less than two years as the Bush presidency ended with a loss to Bill Clinton in 1992. Two years later, Madigan died of lung cancer.

We then come to a period of dynamic changes in Secretary of Agriculture appointments. Bill Clinton chose Mississippi Congressman Mike Espy as his first secretary. Mike was born in Yazoo City, Mississippi, which was also the home of my good friend, country comedian Jerry Clower, "The Mouth of the South." Espy was the first African-American Secretary of Agriculture. He took office on January 22, 1993 and resigned about two years

My longtime camera crew with Secretary of Agriculture Ann Veneman.
FROM LEFT: *Ryan Ruh, Angelo Lazarra, the Secretary, me and Phil Reid.*

later. Three years after he left office, he was indicted and tried for taking gifts of airplane rides, hotel rooms and football tickets while he was secretary. He was acquitted on all charges.

The next Secretary of Agriculture appointed by President Clinton, Daniel Glickman, was the first person of Jewish descent to hold that job. He had a great sense of humor and often joked that he might not be invited to many pork producer banquets because he was Jewish. Dan was a nine-term Democratic congressman from Kansas. As I mentioned elsewhere in this book, one of my international trips was with Secretary Glickman. He served until the end of the Clinton administration.

When George W. Bush became President, he kept the string of "firsts" alive. After the first Roman Catholic Secretary of Agriculture, the first African-American Secretary and the first Jewish Secretary, he appointed the first female Secretary, Ann Veneman. Before her appointment, she'd had a long, distinguished career within the USDA and as Secretary of the California Department of Food and Agriculture. She served as USDA Secretary from January 20, 2001 to January 20, 2005, one of the most prosperous periods for American agriculture, but also one of the most tumultuous. Within weeks after she took office, hoof-and-mouth disease broke out in Europe, prompting stricter sanitary measures over here and restrictions on people and animals traveling to the U.S. from Europe. Bird Flu and B.S.E. also showed up under her watch, prompting several actions to strengthen USDA's regulatory oversight and protections. After four years on the job, Veneman left to become the Executive Director of UNICEF.

"W" started his second term in 2005 with Michael Johanns as our 28th Secretary of Agriculture. Mike grew up on an Iowa dairy farm, milked cows as a youngster and described himself as a farmer's son with an intense passion for agriculture. Mike worked to expand foreign markets for U.S. farm products and pushed forward the growth of the renewable fuels industry. He made his mark as an effective secretary, beginning his term with a series of "listening sessions" across the agriculture community of America. Mike was one of the more personable secretaries, always accessible and, maybe because we shared a dairy farming background, we became good friends. He left the secretary's office in 2008 to run for senator from Nebraska. He was elected and still serves.

When Mike left, the President appointed Chuck Conner, Deputy Secretary to be Acting Secretary of Agriculture. Most of us felt that with a

The 30th U.S. Secretary of Agriculture, Tom Vilsack, on stage with Max and me at the 2012 Farm Progress Show in Boone, Iowa.

year to go in the Bush Administration, Chuck Conner would sit in that seat either as acting secretary until the end, or perhaps the President would name him secretary. But not many of us, including me, thought Mr. Bush would go out and select a new person to become secretary.

In yet another surprise, particularly to the folks of North Dakota, Ed Schafer was appointed secretary. Farm broadcaster Mike Hergert of Grand Forks said it came as a complete surprise to him and his listeners.

Schafer is the grandson of Danish immigrants who farmed the North Dakota plains and while his career was mostly in manufacturing, he said living in North Dakota just naturally gave him a good understanding of agricultural issues. He was a strong supporter of agricultural trade, as well as the alternative fuels industry. Schafer moved into the secretary's office at an interesting and challenging time, with orchestrating the successful conclusion of the writing of the 2007 Farm Bill as one of his tasks.

A couple of weeks after Schafer was sworn in, he found himself in the middle of allegations by animal activists that sick, downed cattle were being abused as they were taken to slaughter at a California packing house. The packing house lost its federal contracts and eventually went bankrupt.

In 2008 Barack Obama was elected president and appointed Tom Vilsack as ag secretary. Vilsack may have seemed an unlikely Obama appointee because after two terms as Governor of Iowa he ran against Obama for the Democratic presidential nomination. And after ending his campaign in 2008, he endorsed Hillary Rodham Clinton and became the national co-chair of Clinton's campaign. But, President Obama appointed him anyway and he had an easy time being confirmed, partly because he comes from a major farm state and was supported by a wide range of agricultural groups, including the Corn Refiners Association, the National Grain and Feed Association, the National Farmers Union, the American Farm Bureau Federation, and the Environmental Defense Fund. Vilsack has been an advocate for the development of many varieties of ethanol, saying that we can't rely on corn alone for energy independence. His policy pushes for other sources, such as switchgrass and various types of cellulosic plants.

Along with interviewing all of the above-mentioned secretaries, I had the opportunity to interview previous secretaries. In the mid-1960s, Michigan State University brought together past secretaries of agriculture. The media was not invited to the event but I caught wind of it and found out when it was taking place and what connecting flights some of them were taking from Chicago to get there. I managed to get on a flight with Ezra Taft Benson, mentioned above, and Charlie Brannan, who was Harry Truman's Secretary of Agriculture. While the Dean of the college didn't appreciate seeing me come off the plane with his honored guests, I had already established interview times with them and my foot was in the door.

Charlie Brannan was a Denver native who held a series of legal and administrative positions within the U.S. government beginning in 1935. He became Secretary of Agriculture in 1948, replacing Clinton Anderson, who was also at the M.S.U. gathering. In 1949, Brannan proposed, as part of President Truman's Fair Deal program, a guarantee for farmers' income, while letting the free market forces determine the prices of commodities. The GOP controlled Congress and the plan died. After leaving office, Charlie Brannan became general counsel for the National Farmers Union. He died in 1992 at the age of 89.

I also interviewed Clinton Anderson, a South Dakota boy who wound up in New Mexico because he'd developed tuberculosis as a teenager and was sent to a sanitarium in Albuquerque. He was a three-term congressman from New Mexico when President Truman tapped him to be agriculture secretary. Anderson served from June 1945 to May 1948. He recalled that one of his most pressing concerns was getting the U.S. agricultural economy propped up. It had been so focused on supporting the World War II effort, when the war ended, the military's demand for farm products dropped off. Anderson helped develop policies on price controls and subsidies, and is credited with incorporating all existing food and agricultural activities under the control of the USDA. He also helped deal with a post-war worldwide food crisis, working with former President Herbert Hoover on Truman's Famine Emergency Committee. By 1948, the worldwide and domestic food supplies had been stabilized and Anderson decided to retire from the Cabinet. In 1949 he became a U.S. Senator for New Mexico, an office he held until 1973. He passed away in 1975 at the age of 80.

Before Clinton Anderson was Claude Wickard, a Hoosier. FDR's ag secretary from 1940-1945, Wickard was born on his family farm near Camden, Indiana. He graduated from Purdue University in 1915. He was Undersecretary of Agriculture when Secretary Henry A. Wallace resigned in 1940 to run for Vice President of the United States. Wickard moved up and during World War II headed the War Foods Administration, promoting increased farm production as a matter of patriotism. He appealed to farmers with the slogan, "Food Will Win the War and Write the Peace".

In 1945, Wickard resigned to run the REA, the Rural Electrification Administration. In 1967, just a few years after I interviewed him, Claude Wickard was killed in an automobile accident on April 29, 1967, at the age of 74.

Also during that Michigan event, I had the privilege of interviewing Henry Wallace, who was a brilliant man. Born on a farm in Orient, Iowa, Wallace was the eleventh Secretary of Agriculture from March of 1933 through September of 1940. His father, also Henry, was the seventh Secretary of Agriculture under Presidents Harding and Coolidge. The younger Henry attended Iowa State University and was very interested in botany and developing new varieties of corn. In 1926, he started the Hi-Bred Corn Company which later became Pioneer Hi-Bred. Dupont bought Pioneer Hi-Bred in 1999 for a reported $10 billion. Mr. Wallace was long gone by then, having died in 1965 of Lou Gehrig's Disease. Three generations of the Wallace family owned and operated *Wallace's Farmer*, a newspaper that kept track of the changes in agriculture and provided news and information to help Iowa and Midwest farmers be more productive. It was started by his father, Henry Cantwell Wallace, continued by him, Henry Agard Wallace and finally, by his son, Henry Browne Wallace. *Wallace's Farmer* became part of the Farm Progress media family.

Secretaries of Agriculture Henry A. Wallace, Claude Wickard, Clinton Anderson, Charlie Brannan and Ezra Taft Benson; it was a fascinating time interviewing those men.

Ole and Lena were in court getting a divorce. The judge looked down from his bench and said, "Ole, I have decided to give Lena $400 a month in support."

"Vell, dat's fine vid me," said Ole. "And vunce in a vile, I'll try to chip in a few bucks myself."

"Big O" with "Big Jim" Thompson, Illinois' 37th, and longest serving Governor, from 1977-1991.

Chapter Ten

Politics and Politicians

"Once you become the secretary, half the people like you and half the people hate you." That was the response Earl Butz gave me in urging me to decline any invitation to become the United States Secretary of Agriculture. My dad's response was a bit different. When my sister Norma told Dad in 1980 that I was being considered by President-elect Ronald Reagan to be his secretary of agriculture, Dad said, "Boy, he'd better not take that job, he'll just become a crook like all the rest of them."

Dad didn't need to worry. I was pretty far down the list of potential nominees — seventh, I recall — but it made me give serious thought to what my answer would have been had the question come. I would have enjoyed being Secretary of Agriculture because I had the opportunity to travel with Secretaries on five trade missions to foreign countries and I think from that standpoint, as an ambassador of American agriculture, I would have been pretty good. I knew enough to hire good people to formulate policy and all that sort of thing and I would have been what Earl Butz was, a real public relations person for agriculture. Keep in mind that only two percent of our population produces all of the food that we eat. They have a very small voice and they need a bigger one. But, many people, in addition to Earl, suggested that I could continue to have an even bigger impact on the agricultural community doing what I was doing. As a farm reporter, I had access to everyone in Washington. They knew I had a big audience on WGN Radio and through our television shows, so I was always welcomed into offices on both sides of the aisle. John Block, an Illinois hog farmer who had been the Illinois Director of Agriculture, became Secretary under Mr. Reagan and I couldn't have been happier. He did an excellent job.

Politics have always intrigued me and had the right opportunity come along at the right time, I might have taken the leap. There was an interesting, actually fascinating, five days in 2004 when I almost leapt. It was when Barack Obama was running for the U.S. Senate. The candidate who had won the Republican primary had to leave the race because of a personal issue and that left the GOP scrambling to find a candidate. In early August, former Bears' coach, Mike Ditka, declined and that's when my life took a turn.

It was the first Thursday of August and I'm sitting in my office when I received a call from a reporter at WLS-TV, Channel 7, asking for an interview about possibly becoming the Republican candidate. He said they were hearing rumblings that I was being mentioned as a possible replacement. It was the first I had heard of it and told them there was no story there, I hadn't been contacted. Then a reporter for WBBM-TV, Channel 2 called and told me that I was being mentioned in downstate newspapers as a possible contender and they wanted to interview me.

Suddenly, people were calling and e-mailing from all over the state, offering to hold fundraisers or contribute to my campaign. Speaker of the House Denny Hastert's office called and invited me to a Sunday meeting at his home near Yorkville, about 55 miles west of Chicago. Denny was there, along with his campaign treasurer and the treasurer of the Illinois Republican party. Denny's wife was there, too, as was my wife, Gloria, who would have a big influence on whether I would run, as you'll read shortly.

This was rapidly becoming very serious. They asked me about my positions, and I admitted that I had some views that wouldn't be popular with the GOP's conservative base, including my stances on gun control and abortion. I'm all for hunting, but I honestly don't believe hunters need automatic weapons. Denny conceded that a few would take issue with me on that, but that he could live with it. Then I told him that I really thought a woman should have the right to choose what happens with her own body, which is a far different view from many Republicans. I added that I have two children, both adopted, and I wouldn't have been so blessed if their mothers had chosen abortion. He seemed to think that was a reasonable position to take and that we could work around it.

The conversation then turned to financing the campaign as I expressed my concern over money. "I am not a wealthy person," I told them. "I cannot afford to come out of a campaign with debt, so how do we pay for it?" Speaker Hastert said, "Well, let's look at where we stand. Most candidates spend up to $4 million to establish name recognition, but you won't have to because you are known throughout Illinois. The primary is history, so you won't have to pay for a primary campaign. And the election is just 90 days away. I think we can do this for $10 million, and I'll get you the first million."

I did somewhat of a double take and said, "Well, that still leaves $9 million to raise. Where do we get that?" Both the Speaker and the state GOP treasurer assured me that the dollars would come in from Republican voters and that I would not be in debt at the end of the campaign.

When we left the Hastert house that afternoon for the drive home, I thought the Speaker had decided I was the candidate, I was ready to accept the challenge and Gloria had reluctantly agreed that I could do it. Her reluctance was mostly based on the campaign coming just a few months after flesh-eating bacteria invaded my throat and nearly killed me, followed by a

Governor George Ryan was a great friend and promoter of Illinois agriculture. Here, we've brought some Illinois corn to Navy Pier in Chicago.

blood clot that came close to finishing the job. I told everyone that the final decision would have to be made by my doctor.

The next morning, I called Dr. Wolfe, who treated the blood clot. Before I could even ask him about running, he barked, "Samuelson, what in the hell are you thinking?"

I said, "Doctor, is that a 'no?'"

He said, "Well, your blood clot would have killed nine out of ten people and you might be able to survive 90 days on the campaign trail, but I don't want you to take the chance."

I wondered how he knew what I was going to ask him. Later, Gloria told me that she called the doctor ahead of time and tipped him off.

That ended my never-started political career. By the way, the eventual candidate the GOP found to run against Mr. Obama was Alan Keyes, who wasn't even an Illinois resident. Many people had the same sarcastic comment: "Thanks a lot, Orion."

GEORGE and LURA LYNN RYAN

Governor Ryan beat me into this life by 35 days. He was born February 24, 1934, in Maquoketa, Iowa. George was the 39th governor of Illinois from 1999 through 2003. He received worldwide attention for his 1999 moratorium on executions in Illinois and, in 2003, for commuting more than 160 death sentences to life sentences. Beyond that, he was an enthusiastic supporter and promoter of agriculture, helping create markets for Illinois farmers worldwide. As of this writing in 2012, George is serving the final year of a prison sentence, having been convicted of federal corruption charges after leaving office.

Lura Lynn will go down as one of the best First Ladies the State of Illinois has had. Partnering with George, her high school sweetheart, she promoted Illinois near and far. She was on the Cuban trade trip with us in 1999 as the governor tried to re-open the door for Illinois products to be sold in Cuba. She also traveled with her husband to South Africa, where they met with Nelson Mandela and opened a trade office.

Lura Lynn tirelessly promoted charitable causes and led development of the Lincoln Library in Springfield. Lura Lynn also promoted literacy programs and after-school initiatives aimed at curbing drug use and crime among young people.

After a long battle with cancer, she died in 2011, not realizing her wish of seeing her husband of 55 years freed from prison.

Lura Lynn Ryan is shown above with a group of 4-H kids involved in one of the many agricultural projects she promoted.

Alan J. Dixon was U.S. Senator from 1981-1993. Prior to that, he was the Illinois Secretary of State and before that, the Illinois State Treasurer.

Jim Edgar was a two-term Governor of Illinois from 1991-1999 and for ten years prior to that was Illinois Secretary of State. He left political life as one of the most well-liked public office holders in state history. BELOW: *Jim and his lovely wife Brenda join me for a photo at the 1998 Illinois State Fair.*

ABOVE: *Talking ag issues with U.S. Senator Richard Lugar at the Indiana State Fair, 1997.* BELOW: *In Washington, DC with Kansas Senator Bob Dole.*

Wisconsin Governor Tommy Thompson, above and below, was always a great interview. Tommy served four terms before going to Washington, DC to become Secretary of Health and Human Services during the first term of George W. Bush.

ABOVE: *President George H.W. Bush was a guest at two Farm Progress Shows. He had a great interest in and appreciation for American agriculture.*
BELOW: *Behind me at the Illinois State Fair are then State Treasurer Judy Baar Topinka, Director of Agriculture Joe Hampton and Governor George Ryan, 2001.*

an evening with
Ann-Margret
Trinity Lutheran Church
Monday—June 11, 1984
The Orrington Hotel
Evanston, Illinois

Patiently, perhaps eagerly, waiting for U.S. Senator Charles Percy to finish talking with Ann-Margret Olsson, a member of Evanston Lutheran Church in her childhood. She came back from Hollywood to lend her name and talents in support of our church.

Conrad Burns was a Missouri farm boy who took a trip to Montana after he got out of the service, met a girl, married and spent many years there as a farm broadcaster. In 1988, he won a seat in the U.S. Senate and served 18 years.

Ole died. Lena went to the local paper to put a notice in the obituaries. The newspaper clerk, after offering her condolences, asked Lena what she would like to say about Ole.

Lena thought about it a moment and said, "Vell, yust say 'Ole died.'" The clerk, surprised by the brevity, said, "That's it? Just 'Ole died?' There must be something more you'd like to say about your husband. If you're concerned about the cost, the first five words are free. His obituary must say something more."

Lena thought for a few more moments. "Ya, sure. You put, 'Ole died. Boat for sale.'"

The world's agricultural commodity prices are set each day on LaSalle Street in Chicago. This photo, taken by the Chicago Tribune's Carl Wagner on April 4, 1998, marked a milestone in the life of the Chicago Board of Trade. From left, CBOT Chairman Patrick Arbor, Mayor Richard M. Daley, CBOT President Thomas Donovan and I ring the opening bell as the Board of Trade celebrated 150 years of trading.

Chapter Eleven

For Crying Out Loud

"Oh no you're not!" was my response when my son David came to me many years ago after he finished college and said, "Dad, I'm going to be a trader at the Chicago Board of Trade." Since I started covering the CBOT in 1960, I had seen too many friends get worn out by the trading business. They developed ulcers, they became alcoholics, they ruined their marriages, they died of heart attacks — I didn't want him to be a trader. "David," I advised, "you don't have the discipline to be a trader. Don't do it." Well, he'd already been bitten by the bug after working summers during high school and college as a runner and a clerk on the Merc's adrenaline fueled trading floor. David ignored my advice and has proved me wrong with a long, successful career as a trader and none of the resulting ailments that had me worried. Of course, I take credit for his success because I told him he couldn't do it. He took that as a challenge and has done very well.

One key to David's success is his discipline. If he gets into a market position and he sees signs that it's going against him, he's out. He'll take a small loss but get out before it becomes a big loss. And if it starts going his way, he's already determined what price he wants to see, and if it gets close to that level, he takes it.

In coffee shops all across rural America, where many of the world's problems are solved on a daily basis, farmers routinely blame traders like David, "speculators in silk suits," for gobbling up profits from grain sales that should go to farmers who raised the grain, not to people sitting at a computer, never getting their hands dirty. What those gripers need to understand is that without people like David, there would be no market for their crops.

Traders — speculators if you prefer — provide liquidity to the market. Liquidity means having cash in the market, and the people who supply that cash are the speculators. That allows the farmer to sell corn at eight dollars a bushel if he wants to because there is a trader on the other side that has the liquidity, the cash, to take that opposite position, and that's what makes a market. You can't have just sellers; you can't have just buyers. You've got to have both sides of that market covered, and that's liquidity.

Too many farmers don't have the discipline to make good decisions in trading, even though they have an advantage over David. They have the product. He doesn't. All he's got is a piece of paper that says he's

During a live broadcast from a Beef Cook-Off in Pioneer Court (next to the Tribune Tower) with execs from the CME and CBOT.

long wheat or he's short wheat. But, when talking to corn farmers, I've often heard them tell me, "Boy, when corn gets to 'x' dollars a bushel, I'm selling it." Well, corn would get to 'x' dollars and nudge a little higher, but when I called them up and said, "Hey, congratulations, you got your price," they'd say, "Oh, I didn't sell it, Orion." "Why not?" "Well, it's going higher, so I'm hanging on." And you know how that story turns out.

It seems farmers rarely sell on an up market, but boy, do they ever sell on a down market, because they keep thinking the price is going to go up, and if they sell and it goes up another ten cents, they've lost that ten cents. So, they say, "It's going my way. I'll ride with it up." And then two days later, the market will start to go down, and they'll think, "Well, I'm still alright. It'll come back tomorrow." And then the second day, it goes down again, and by the third day, now, they're nervous and start selling, and everybody else in that same boat starts selling, and it's just down, down, down. That's where discipline comes in. Producers need to know their cost of production, determine what price will give them enough profit and then call their trader and put their order in to sell at that level. When it gets there, they have to have the discipline to sell it and not look back.

Some of the best market advice I ever heard came from a trader on the floor of the Chicago Board of Trade, Uncle Julius — a colorful little fellow, a pure speculator who died at the age of 84 and left everything to pet hospitals — made millions and lost millions during his 54-year career. Every time I'd go onto the trading floor, Uncle Julius would come up to me and tell me the same thing he told me dozens of times before.

"Orion, you've got to tell your farmers some things about how they market. First of all, tell them to quit trying to sell at the top of the market. Top is here and gone three weeks before you know it was the top. Secondly, if you think corn is going to three dollars, put your order in at $2.97, because everybody else will probably be at three dollars, and so the market will hit three, and everybody sells, and it turns around and goes down. Leave a little for the other guy, and you'll go to the bank." And his third one that summed up the first two was, "Remember, in the market, there's always profit for the bulls, always profit for the bears, never any profit for the hogs — they get slaughtered!"

Farmers have been getting wiser over the years, but it has also gotten more complicated. Volatility, created by the near complete conversion to computerized trading, has led to huge swings in the prices of corn, soybeans and wheat. For years, there would rarely be days when the prices moved more than a couple of cents per bushel and potential gains and losses were mostly moderate. But with today's wild swings, millions of dollars can be gained or lost in seconds. Back in the sixties, you looked at supply, demand and the weather and that's pretty much all traders had to consider. But the trading atmosphere changed when the funds began bundling commodities and tying corn to oil to copper and using those in investment portfolios for investors like you and me if we're into a mutual fund. That totally changed the complexion. And when ethanol and biofuels entered the picture, the price of oil affected the price of corn and soybeans, and particularly soybean oil, because that goes into biofuels. So, if you're wondering which way the soybean market is going, you look at the price of crude oil. Never had to do that before.

Another factor that made trading more complicated was when the options market was added. It was needed to address the issue of basis, where the basis price on the river could be twenty to thirty cents off from the futures price at the Board of Trade. The options market was developed so that puts and calls were added as tools. This is why I'm glad to see these young farming people going off to college and coming back to the farm armed with the latest knowledge about the markets so that they can really protect themselves and their farms in the marketplace much better.

The role of "chief marketing officer" on many farms has been increasingly taken over by women. As I like to tell the men in the audience during my speeches, "I know why that's happening. It's because you guys sit in the combine, and you see that corn and soybeans coming out into the wagon, and you think, 'Man, that's the prettiest stuff I've ever grown. I love it,' and you don't want to sell it. But when your Chief Marketing Officer knows there's a bill to pay or a pickup to buy, she sells it."

I talk often about a farming couple in Northern Illinois who are friends of mine. They milk 250 cows and grow 1,500 acres of cash corn and soybeans. He oversees the majority of the field and dairy work and she does all the marketing. Every day, she sits at her computer for an hour or so, but here's what separates her from a lot of producers: her goal is not to sell at the top of the market. Her goal is, at the end of the market year, to have sold everything in the *top third* of the market. She's done it that way for many years and that philosophy has made a tremendous difference for them. She markets milk in the futures market, and she sells the grain.

The trading floors at the Chicago Board of Trade and the Chicago Mercantile Exchange have provided me with many great memories. Before open outcry trading became nearly extinct, the pits were filled with some of the most outrageous and colorful characters you could imagine. The types of men I was afraid my son David would become tended to live life on the edge, feasting on the adrenaline swings that came with their wins and losses. Dressed in bright-colored homemade vests designed to be easily picked out of a crowd, they yelled at the top of their lungs, flashing hand signals to indicate whether they wanted to buy or sell and in what quantities. Just like Uncle Julius, they made and lost money on a daily basis. There aren't many left and as much as some people, including me, want to keep open outcry trading alive, I'm afraid it's just a matter of time before the pits will be silent.

A footnote to the market story... Over the years I've had many producers tell me why they don't like the futures markets, but let me share the most unusual letter I received. It happened in the '70s when grain prices went through a brief time of sharp price declines for no apparent reason to producers and a farmer in Iowa wrote two pages on notebook paper telling me why the markets hurt farmers and should be closed. He presented his story very well and then on page three, he thanked me for the information I presented every day and closed his letter with a request: Could I please start quoting the December corn price two months earlier? I guess he didn't realize that if there was no futures market, there would be no December corn price for me to quote.

Ole bought a new car and decided to take Lena for a drive. As they were cruising through town, a policeman pulled them over and told Ole that he was doing 50 miles an hour in a 30 mph zone. "Oh, no!" Ole protested. "I vas only doing tirty, officer."

"No, sir, you were doing fifty," replied the cop.

"Really, officer, I vas only doing tirty," Ole replied raising his voice.

"Well," bellowed the cop, "I clocked you doing FIFTY!"

"TIRTY!" yelled Ole.

"FIFTY!" yelled the cop even louder.

At that point, Lena, sitting in the back seat and trying to be helpful, spoke up.

"Officer, you really shouldn't argue vit Ole ven he's been drinking."

Peder had been a huge Green Bay Packers fan for decades but had never been to Lambeau Field to see his heroes play. On a cold, sunny Sunday afternoon when the Packers and their bitter rival Chicago Bears were playing, he decided to go to the field to see if he could get a ticket. Alas, they were sold out, but as he was standing by the Vince Lombardi statue, a stranger walked by and offered him an extra ticket. Peder was ecstatic! When he climbed to his seat, he found it was in the uppermost row in the corner of an end zone. Peder didn't mind, after all, he was at the Packers-Bears game. But, as he scanned the stadium with his binoculars, he noticed a guy sitting about 10 rows up from the 50-yard line with a vacant seat beside him. Looking closer, he sees that the man is an old friend, Ole, whom he hadn't seen for years. The seat remained empty through the first two quarters, so at half-time, Peder walked down to say hello to his old friend, and asked Ole why he had a vacant seat in such a choice location.

Ole says, "My wife, Lena, and I bought deese seats a long time ago. But, ya know, Lena passed away."

"Oh, I'm really sorry to hear dat," Peder said, "but why didn't you yust give da ticket to another relative or a friend?"

Ole replied, "Well, dey are all at her funeral."

Ole and Lena are at the kitchen table in their Chicago home. One snowy morning as they were having their coffee and listening to WGN Radio, the announcer said, "Tom Skilling's forecast says there will be 10 inches of snow today and a snow emergency has been declared. To make sure the snow gets removed properly you must park your car on the odd-numbered side of the streets." Ole gets up from his coffee and goes out and moves his car to the odd-numbered side of the street.

The next day, as they're having their cups of morning coffee, the WGN announcer says, "There will be eight inches of snow today and a snow emergency has been extended. Today, cars must be parked on the even-numbered side of the streets." So Ole heads outside to park the car on the even side of the street.

A few days later, another storm moves in and as Ole and Lena sit at the kitchen table with their morning coffee, the WGN announcer says, "We are expecting five more inches of snow today and the snow emergency has again been issued. You must park your car on the..." And just then, the power goes out. Ole and Lena don't get to hear the rest of the instructions. Ole says, "What are we going to do now, Lena?"

Lena says, "Aw, Ole, dis is nuts! Yust leave da car in da garage."

It's always a pleasure for me to talk about my 4-H roots.

Chapter Twelve

4-H and FFA

Occasionally, I'll think back to my preteen years when I became a member of the O'Connell Rustlers 4-H Club and just how important that organization was to me. I didn't realize how big an impact it was going to have on my life because I was busy having fun! Attending meetings, going on trips and just being involved with other kids in a club setting was, first of all, very entertaining. Learning how to keep records and how to raise, groom and show my cows, and what made one chicken more desirable than another was a bonus. And other aspects of being a 4-H club member have had a lasting impact on my life. What I learned from competing in 4-H public speaking contests helped me become an effective communicator. Each time I stood before the judges and gave my speeches, I was graded on things like stage presence, enunciation, poise, posture and how to express myself in ways that would hold the audience's attention. Each time, I learned and each time, I got better.

Head, Heart, Hands and Health are the four "Hs" that nearly every farm kid — and millions of city kids — have grown up with since the turn of the 20th century, when 4-H was formed by the USDA. But it wasn't created just to give country kids something to do. The USDA was having trouble getting farmers to accept new ways of raising crops and livestock that researchers at land-grant universities were discovering. The USDA found that if these new methods were shown to young people, they would then share them with their parents. So 4-H became a conduit for introducing new agriculture technology to the adults. I don't recall whether I passed along any new knowledge to my dad, but my projects, besides public speaking, were dairy and poultry.

Each of our meetings began with a recitation of the 4-H Pledge, and I'll bet there are millions of adults who can still recite at least some of that pledge that's buried in their memories:

As a true 4-H member, I pledge...

My head to clearer thinking,
My heart to greater loyalty,
My hands to larger service,
My health to better living
for my club, my community,
my country and my world.

The lessons taught in 4-H continued when I got into high school and joined the Future Farmers of America. Of course, lessons are often delivered on the back of disappointment. When I thought I would win an FFA statewide speaking contest, I came in fourth out of five. My speech topic was, "Will Unions Work in Agriculture?" That was a rather controversial subject at the time. One of the judges didn't care for my approach and took me apart in the question and answer session.

My project work didn't go much better. Each year, the Lions Club in our town would present two members of the FFA an already bred, purebred Berkshire gilt; in other words a pregnant sow that would be delivering a litter of pigs. The Lions Club chose me as a recipient, the gilt was delivered and I took great care of that hog. The time came for the litter to be due, but the Pork Stork didn't fly. The farmer who donated the gilt guaranteed that she would deliver, so he brought in another one and, again, I took tender loving care of that animal. When it should have been time for her to deliver, the Pork Stork, again, didn't fly.

We had Guernsey dairy cattle and we decided that for my next project, I would raise a registered Guernsey heifer. She was a beautiful animal with the prettiest eyes you'd ever see in a cow. She, too, turned out to be barren and we sent her off to market. So that was three failures and the next year I took on corn as a project and didn't have any breeding problems there. As I

In the spring of 1950, FFA speaking contest winners from several Midwestern states were brought to Chicago as guests of WLS Radio. I'm in the back row, seventh from the left.

Photo courtesy Dick Resler, Bourbonnais, Illinois

Photo courtesy Dick Resler, Bourbonnais, Illinois

Our WLS Radio hosts introduced us hungry farm boys to Chinese food during our 1950 trip to Chicago. I'm third from the left, focused on making my chopsticks work.

say when I tell that story in my speeches, there must have been something about the water in the Kickapoo Valley where I grew up, because both of my children are adopted.

If you don't mind me getting on my soapbox for a moment, it bothers me that the showcases for 4-H and FFA members, county and state fairs, are frequently mentioned as targets of budget cuts.

There is no question the economic crisis is causing major challenges for governors and state legislatures across the nation, forcing them to take a look at long-standing events and institutions, seeing if there is a way they can save a dollar or two. A few years back, a headline in the *Detroit News* caught my attention. It read, "State Fair in Jeopardy." The opening paragraph stated, "Governor Jennifer Granholm wants to eliminate state funding for

With the officers of the Illinois FFA, 2002.

the 160-year-old Michigan State Fair. Her short-term reform plan would end all state financing of the Michigan State Fair in Detroit, the nation's oldest, as well as the Upper Peninsula State Fair in 2010. The state fair — a showcase for Michigan's $64 billion agricultural industry — is an agency of

Hosting 4-H and FFA members on a WGN Noon Show remote.

the Department of Management and Budget, and is held on state property." The story went on, "It's supposed to pay for itself, but the state treasury has been covering shortfalls ranging from $50,000 to $1.3 million in recent years. Attendance has dropped 39% since 2000 and corporate support has also declined."

This problem is not unique to Michigan. I've heard from people in other states who are concerned about possible cutbacks in county and state fair funding that in many cases could lead to the closing of their fairs.

It's no secret how I feel about county and state fairs. They are important to agriculture as showcases and as opportunities to educate city fairgoers. They provide the opportunities for 4-H and FFA members to demonstrate their accomplishments in project work and compete for top honors in the show ring.

If attendance and revenues are declining at your county or state fair, the Michigan story makes it very clear you will get little or no financial help from government. That puts the responsibility on our shoulders to generate new funds and develop new ideas to attract more people so we can keep county and state fairs off the Endangered Species list.

It is a wake-up call for us who want our kids and grand kids to be able to enjoy fairs as we did when we were growing up. Times are challenging, but if we work together, I'm confident we can find ways to save and grow this important slice of our American rural heritage.

Another concern of mine, since so few of us are left on farms and doing the work of putting food on the world's tables, is where the new voices will come from. Over the course of a year, I attend many meetings and conventions of national and state farm organizations and commodity groups. When the current leadership says it's time to move on or retire, we will need young, new, capable leadership to take over. It should be a concern for all of us in agriculture because in this highly technological world of communications, we need men and women who can lead and deliver an articulate message on behalf of America's farmers and ranchers. But, as I have in each October since 1958, I spend a week at the National FFA Convention and my concern is greatly diminished. More than 50,000 of the nation's brightest young men and women, dressed in their crisp National Blue corduroy jackets with Corn Gold trim gather to talk about their and agriculture's future. You can't help but be enthused when you talk to these impressive young people who, at a very young age, have set lifetime goals and in many cases, have established the pathways to reach those goals.

Keep in mind that the FFA of today is a bit different than it was when I was growing up. With the change in our lifestyle and our world, the organization has adapted, including changing its name from Future Farmers of America. It is now officially known as The National FFA Organization, because with more than 300 career opportunities in the world of agriculture and agribusiness, many of these members will not become farmers. You will now find urban chapters; in fact, the nation's largest FFA chapter is at the Chicago High School for Agricultural Sciences. Still, the majority of members nationwide come from high schools in rural communities where agriculture is the leading industry. There was also a time when young ladies were not allowed to be members of the FFA, which never seemed right to me. I remember back in the '60s, going on the air year after year saying, "Well, at the FFA convention again this year, the boys voted against the girls." Today nearly half of FFA members are female and women hold about half of state leadership positions.

I also remember when the membership of the FFA consisted of all white students and African-American students had a separate organization, the New Farmers of America. In the '60s, the two groups merged and today, about 75% of FFA membership is Caucasian, 16% is Hispanic, 4% is

Some of the largest 4-H and FFA chapters in the country are in Cook County, Illinois. Until he retired in 2011, Dr. James Oliver was Assistant Dean of the College of A.C.E.S. at the University of Illinois and oversaw the Extension Service outreach in Cook County.

African-American and 2% is Native American.

FFA has expanded the leadership building program beyond farms and ranches and moved it into urban and suburban communities. About 70% of FFA members still live on farms, but 10% live in urban and suburban areas and 20% live in small towns. FFA itself has moved. Its headquarters left Alexandria, Virginia and re-located to Indianapolis, Indiana. And, in a very controversial move, the FFA Convention, held for decades in Kansas City, left and now alternates between Louisville and Indianapolis. That departure from Kansas City resulted in some hard feelings being left behind. That anger led to the formation of the Agriculture Future of America. It has grown over the years into a very positive organization to help college students who are studying for careers in agriculture to come together and network, and to meet agribusiness executives who may someday be their employers. Hundreds of AFA college students from nearly 40 states meet each year in Kansas City and I've had the pleasure of attending these gatherings and, again, there is no doubt in my mind that the future of world agriculture is in solid hands with these students who are excited and ready to be our next generation of leaders.

But, back to the FFA; as agricultural technology has changed, so has the agricultural curriculum, with heavy emphasis on science. Most agricultural education programs offer agriscience, advanced agriscience and biotechnology, agricultural mechanics and horticulture, and a growing number offer animal science and environmental courses.

These young people have great ideas on what they want to do to improve their communities and the world. An important part of the FFA program is community service. That's why, for two days during the convention in Indianapolis, FFA members participate in the "Days of Service" program, building homes for Habitat for Humanity, packing food baskets at food pantries and repairing and cleaning city parks and playgrounds so that when they leave, the residents of the Circle City will remember what their visitors did to improve their city.

Each week on our national television show, Max and I salute an FFA chapter and, in recent years, I have hosted the RFD-TV live coverage during the four days of the national convention. It gives me the opportunity to interview many of these impressive young men and women, including the four annual winners of the American Star Awards. Those four "Stars Over America" represent the best of the best among thousands of American FFA

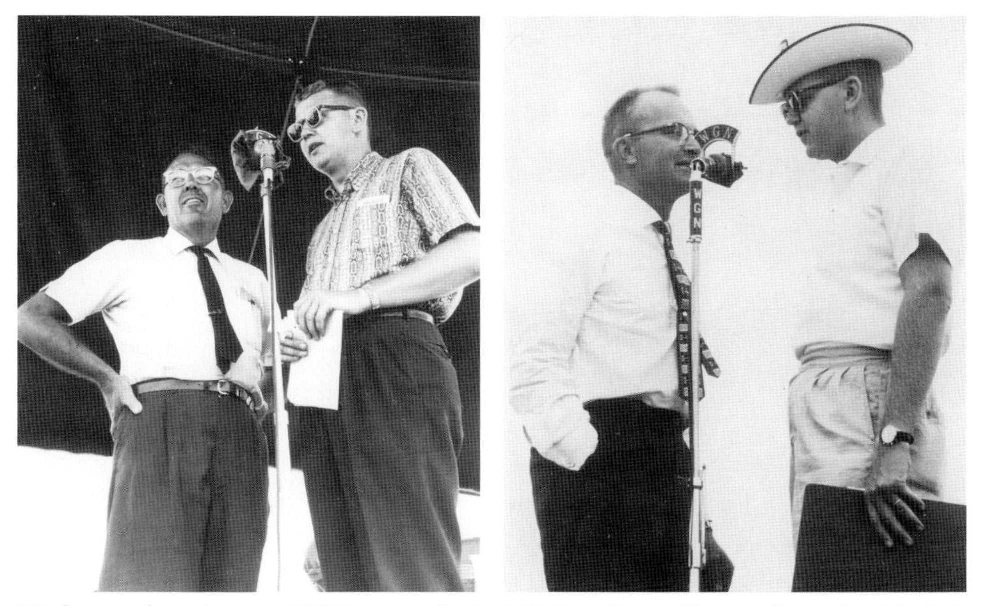

My first words on the air at WGN were at the M & W Farm Power Show on September 26, 1960: "This is WGN Radio, clear channel radio, serving the nation from Chicago."
ABOVE LEFT: *M & W Gear Company President Elmo Meiners.* RIGHT: *M & W Vice-President Tony Munzell. M & W, of Gibson City, Illinois, hosted the Farm Power Show, an annual outdoor farm show in the '50s & '60s. WGN broadcast live from the show all three days each year.*

Quite often, our remote broadcasts didn't include a stage and a band. We were in farm fields where we examined test plots to help farmers stay abreast of the latest innovations in seeds, fertilizers, herbicides and pesticides.

The WGN Barn Dance, circa 1964.

WGN Radio

WGN Radio

In the WGN Bradley Place studios' garage with our new 1963 Chevy farm wagon.

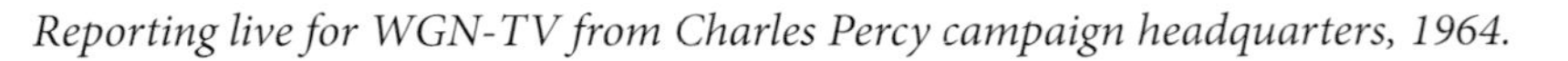

Reporting live for WGN-TV from Charles Percy campaign headquarters, 1964.

WGN Radio

WGN Radio

The WGN Country Fair noon show, circa 1964.

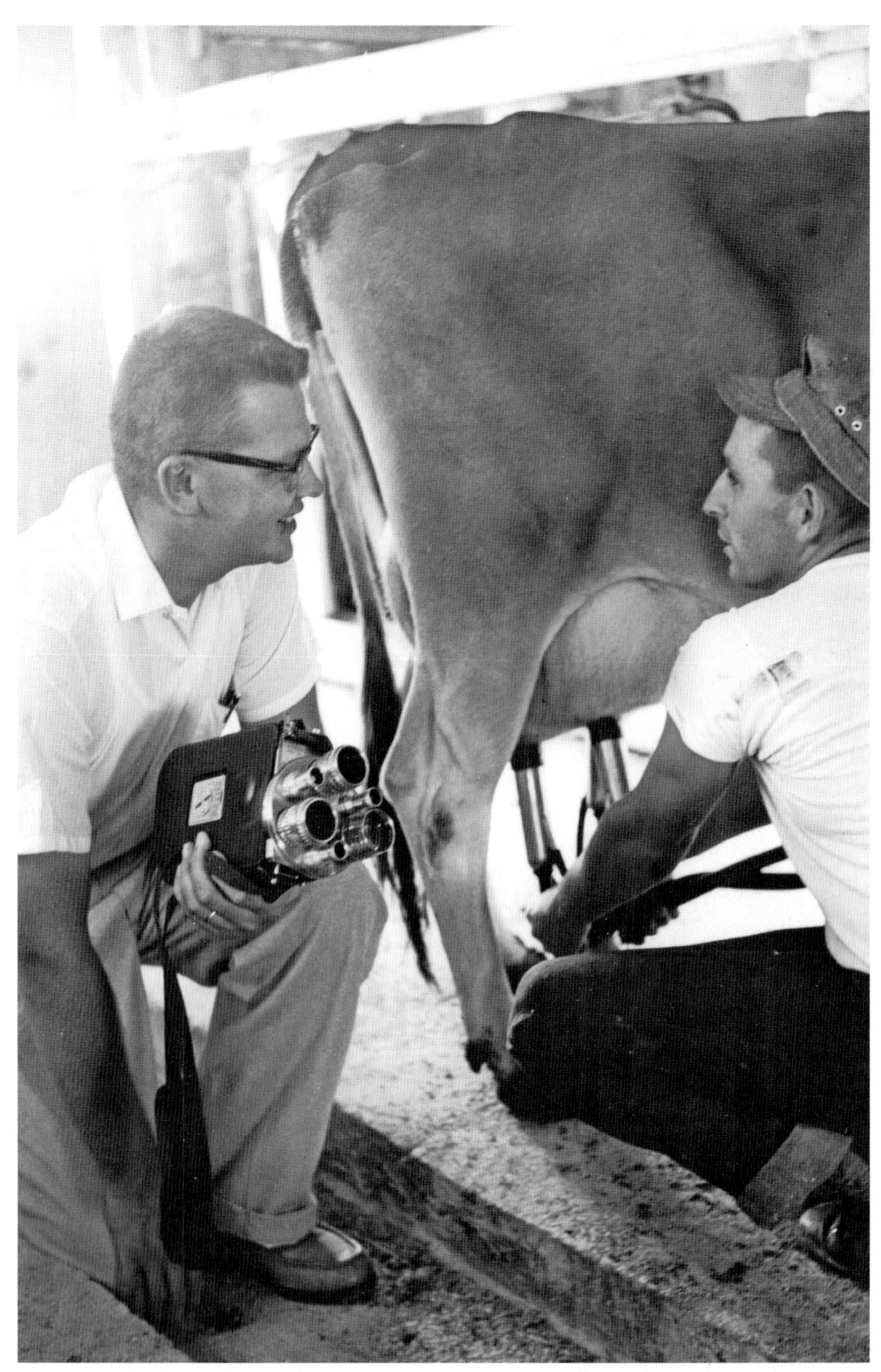

In 1962 with second generation McClean County, Illinois dairyman Gordy Ropp, who served as Illinois Director of Agriculture from 1970 through 1979.

Gordy Ropp and his wife Roberta, shown here with daughter Diana and son Darren were selected by the Pure Milk Association as the 1962 Outstanding Young Dairy Farm couple.

A WGN Radio live broadcast from the Illinois State Fair in the 1970s. Gordy Ropp, the Illinois Director of Agriculture is at the microphone. That's Lino Frigo on accordion.

WGN Radio

On the phone with a producer during a break in a "Milking Time" remote from the International Livestock Exposition with WGN Radio engineer Karl Michaels, mid-1960s.

With Tom Donovan, CEO of the Chicago Board of Trade and William Brodsky, CEO of the Chicago Mercantile Exchange, circa 1995.

WGN Radio

Lola Dee was one of our "Noon Show" featured singers in the 1960s. She's pictured to my left, next to band leader and accordionist Lino Frigo in the older photo and, above, at my WGN 50th anniversary party in September, 2010.

Air Orion and me. INSET: *Los tres amigos at a Scottsdale fiesta; pilot Jerry Lagerloef, me, pilot Phill Wolfe.*

A group of former National Association of Farm Broadcasters presidents, November 1994.
BACK ROW, LEFT TO RIGHT: *Roddy Peoples, Ken Tanner, Evan Slack, me, Johnny Hood, Keith Kirkpatrick, Ron Hayes, Earl Sargent, Rich Hill, George Logan, Taylor Brown, Gene Williams.*
FRONT ROW, LEFT TO RIGHT: *Dix Harper, Ray Wilkinson, Curt Lancaster, Herb Plambeck, Art Sechrest, Lynn Ketelson.*

Radio legends Clark Weber and Lyle Dean are among our many great Chicago friends.
FRONT ROW, LEFT TO RIGHT: *Lyle's late wife Sharon, Gloria, and Joan Weber.* BACK ROW, LEFT TO RIGHT: *Lyle, me and Clark, 2006.*

Paul Wallem is as close to a brother as I'll ever have. We have been great friends for decades. He owned an International Harvester dealership in Belvidere, Illinois and his love of aviation (notice the shirt!) led to our becoming partners in Air Orion. Paul has a multi-talented family, including his daughter, Linda, who was a creator of Showtime's "Nurse Jackie," and his son, Stephen, who is an actor on the show, playing a nurse nicknamed "Thor."

On board the "Heidi Ho," also known to WGN listeners as "The Yacht Lyle Dean," with Lyle and Sharon, and Joyce and Dick Grassfield, at whose home Gloria and I were married, and Christine Collins.

The Samuelson clan, photographed in 2007 in Arizona. LEFT TO RIGHT: *son David, grandson Matthew, granddaughter Grace, daughter Kathryn, me, Gloria and daughter-in-law Carla.*

The newlyweds on November 23, 2001. We were married in Wauconda, Illinois at the home of longtime friends Dick and Joyce Grassfield, officiated by Pastor Joel Benbow of Trinity Lutheran Church of Evanston.

In 2001, I was honored to receive The Order of Lincoln, the State of Illinois' highest award for individual achievement. With me are Ron Warfield and Harold Steele, both past presidents of the Illinois Farm Bureau.
BELOW: *University of Illinois President Robert Easter, Chancellor Phyliss Wise and College of A.C.E.S. Dean Robert Hauser after my 2012 Spring Commencement address.*

degree recipients. Finalists for the award have mastered skills in all of the areas important to agriculture: farming, agribusiness, finance and management and agriscience. I also interview each of the newly-elected leadership teams, and each time I do, I come away feeling very confident about the future leadership in the country and the world. I can assure you the leadership skills of these young people grow stronger every year.

The FFA motto:

Learning to do,
Doing to learn,
Earning to live,
Living to serve.

4-H and FFA are the two best leadership building organizations for young people on the planet today. I was privileged to serve two, three-year terms on the Board of Trustees of the National 4-H Council. I was gratified to see a growing 4-H membership and strong individual and agribusiness financial support for 4-H. Both 4-H and FFA, on state and national levels offer millions of dollars in college scholarships each year to encourage involvement in higher education and agricultural careers.

My personal debt aside, our country owes much to these organizations and I urge you to support them with your time and money. Become a volunteer 4-H Club leader, encourage your school administrators to maintain agricultural education and FFA in your high school. Agriculture needs strong leaders and voices.

Our survival depends on these young people. The agriculture industry is booming, it is one of the few bright spots in our economy today. People need to eat and the number of people in the world grows each day. I've seen predictions that agricultural production will need to increase 70% by the year 2050 in order for the world to have enough food to eat. That means we need more young people to be involved in agriculture and by that, I don't mean just cows and plows. We need farmers, but we also need veterinarians to keep livestock healthy and scientists to continue the great work that has been done in engineering animals and plants so that they produce more, produce it quicker and, in the case of plants, with less water. Seed companies are adding employees by the tens of thousands. Graduates from Midwest agricultural schools are walking into high paying jobs with companies like

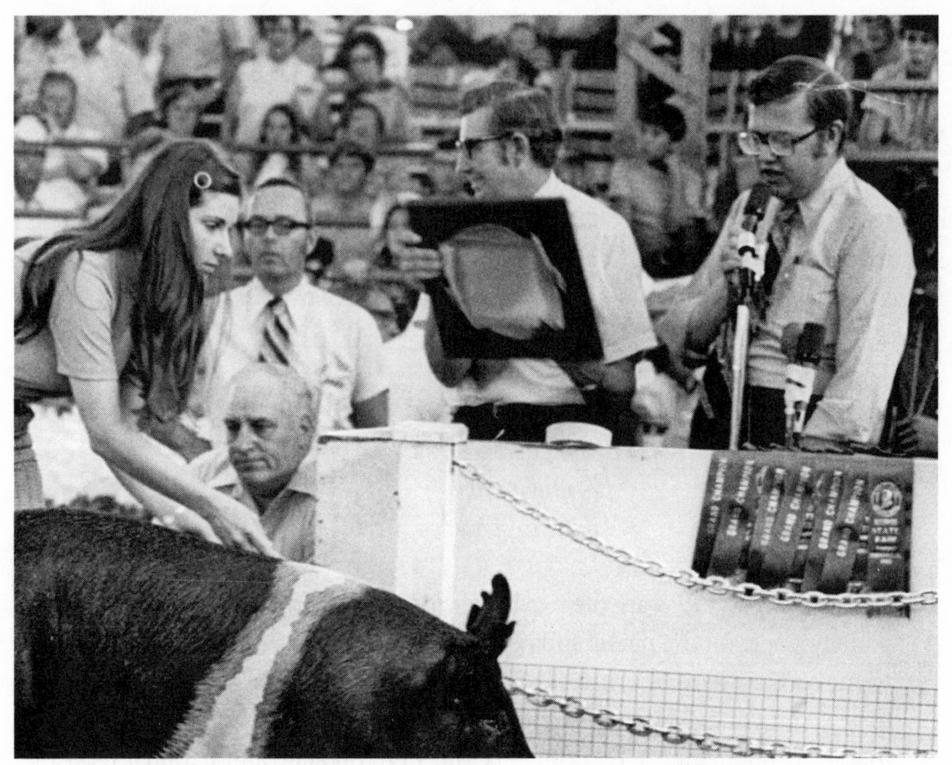

I'm proud to have served for many years as Master of Ceremonies at the Illinois State Fair Sale of Champions.

ADM and Pioneer Hi-Bred. And once the food is grown and processed, there are many other ag-related jobs ranging from dieticians to chefs. As I've said many times, if you eat, you're involved in agriculture. And the two groups doing the best job of preparing young people for vital careers in agriculture are the 4-H and FFA.

I'd also like to talk about a very important part of the FFA program, the Agriculture Teacher/FFA Advisor. The shortage of qualified ag teachers is the greatest challenge facing FFA and agricultural education today. If you are a student looking for a rewarding career with immediate job openings across the country, consider becoming an ag teacher. I speak from experience when I say that you will have the opportunity to change lives, just as my ag teacher did for me in 1951 when he led me to this career in agricultural communications.

I'm very grateful for the recognition given to me by the 4-H and FFA, including the National 4-H Alumni Award, National 4-H Partner Award from USDA, National 4-H Council Award and Honorary FFA American Farmer Degree.

Of course, the need for education doesn't end when a young person's

Uncle Bobby didn't enjoy speaking in public, but was kind enough to say a few words at the dedication of "my" building in 1997. That's Governor Jim Edgar over Bob's right shoulder.

eligibility for FFA ends. The Illinois Agricultural Leadership Program was formed in 1980 with the idea of raising a new bumper crop in Illinois: the state's best and brightest agricultural leaders. It wasn't that we didn't have men and women who were brilliant in all things agricultural and weren't willing to step up and assume leadership roles. We did, but it was evident that they often needed development in other areas critical to leadership such as communication, business, marketing, policy development and trade. A non-profit educational foundation was formed with the mission statement of "developing knowledgeable and effective leaders to become policy and decision makers for the agricultural industry." The Illinois Agricultural Leadership Program, one of several state leadership programs across the country, has been a hugely successful program and I'm privileged to have served on the Board of Directors for its first 28 years.

THE ILLINOIS STATE FAIR

Of the dozens of events I attend each year, few, if any, are as special to me as the Illinois State Fair. Every August for over two-thirds of my life, I've taken my broadcasts to the state fairgrounds in Springfield to join hundreds of thousands of others in celebration of agriculture in the Land of Lincoln.

It is perhaps the most exciting ten days of the year for the young

The State Fair was a perfect place to raise money. This was from a Pork Fundraiser in the late '60s.

The Butter Cow has been a popular attraction in the Dairy Building at the state fair since the early '40s. It's sculpted from about 600 pounds of butter and for more than a quarter-century, Norma Lyon, a dairy farmer's wife and mother of nine from Iowa, was the artist. During a World War II butter shortage, the cow was dismantled and buckets of butter were given to fair employees. In later years, Ms. Lyon recycled the butter, using it in other sculptures as she traveled the Midwest, gaining worldwide fame as "the butter cow lady." She passed away in 2011 at the age of 81.

Illinois State Senator Cecil Partee, Illinois Director of Agriculture John Block, a baby pig and me at the State Fair in 1977.

men and women who have fed, groomed and pampered their show animals for months to have the chance to compete in the show ring against the best of the best. There's nothing to compare with watching the 4-Hers and FFA members leading or guiding their impeccably prepared animals into the ring, hoping to leave with a blue or purple ribbon and maybe even a trophy. And when it's time for the Sale of Champions, which I've emceed for years, the emotions only grow deeper as you watch some of the exhibitors struggle with having to part with animals that have occupied the majority of their time for such a long time. From the moment months earlier when they picked out a calf or other animal hoping that it would develop into a champion, there's an inevitable bonding that takes place. Of course, it is a business and they are being paid for their animals, but it can still be a very emotional moment.

For farmers, ranchers and other producers, the fair, whether it's a state or local fair, is still what the first fairs were designed to be: a chance to renew old acquaintances and compare notes with each other about what's working and what isn't, and about how good or bad the weather has been.

The official function of the State Fair, as stated by the Illinois Department of Agriculture, is to showcase Illinois agriculture and offer wholesome family entertainment. That it does, with great flair and success each year. Visitors not only get the opportunity to see the state's best livestock, there's great musical entertainment and, of course, the carnival midway. There's also food, a lot of it and a lot of different kinds. Long before I arrived on the scene, sometime in the '40s, the corn dog became a staple in the state fair's food booths. Food on a stick continues to be the most popular way to eat food at the state fair and each year, something new is the rage. But in 2011, the food-on-a-stick people may have outdone themselves: deep-fried butter on a stick. Imagine an ice-cold, half-stick of butter, coated in a batter containing cinnamon and sugar, then deep-fried for a few seconds to seal in the melting butter. I'm told it's delicious... sloppy, but delicious.

Ole lived across a river from Clarence. Ole and Clarence didn't like each other and they were all the time yelling insults across the river at each other.

Ole would yell to Clarence, "If I had a vay to cross dis river, I'd come over dere an beat you up good, yeah sure, ya betcha by golly!"

This went on for years. One year, the state built a bridge across the river right by their houses. Ole's wife, Lena, says, "Now is your chance, Ole, vy don'tcha go over dere and beat up dat Clarence like you said you vood?"

Ole says, "By yimminy I tink I vill do yust dat." Ole headed out the door for the bridge, but when he sees a sign on the bridge and stops to read it, he turns around and goes back home.

Lena asked, "Vy did you come back so soon?" Ole said, "Lena, I tink I change my mind bout beatin up dat Clarence. Dey put a sign on da bridge dat says, 'Clarence is 13 ft. 6 in.' He don't look near dat big ven I yell at him from across da river."

Courtesy: Don Peasley Photography

Notice the looks on the faces of Miss Wonder Lake and Miss Crystal Lake.

Chapter Thirteen

Queens, Princesses and Cow-Milking Contests

Occasionally, I was called upon to spend time in the company of pretty young women. There were queens and princesses of all sorts — Dairy, Pork, Rodeo, Miss This and Miss That — I crowned them all and as you can see in the photo on the opposite page, I kissed a few, too. Just part of the job... just part of the job!

Governor Richard Ogilvie crowning Linda Lawyer as Illinois Pork Queen, c.1970.

Courtesy: Don Peasley Photography

ABOVE: *A princess and a queen: The McHenry County Dairy Princess and Miss McHenry County, 1981.*
BELOW: *About 15 years earlier with Miss McHenry County and her court.*

Courtesy: Don Peasley Photography

ABOVE: *Interviewing an Illinois Pork Queen at a Hawaiian-themed fundraiser at the Illinois State Fair.* BELOW: *Miss Sunflower at our "Top O' the Morning" show.*

WGN

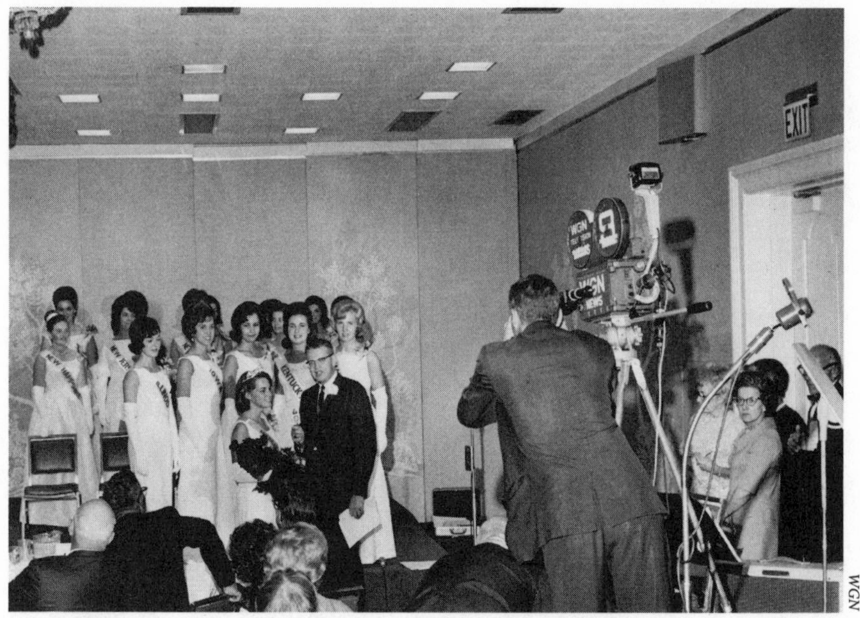

WGN

The American Dairy Princess pageant was sponsored by the American Dairy Association from 1955 to 1972. All but four of the pageants were in the Chicago area. Each year's winner traveled across the country promoting milk and milk products as a full-time employee of the American Dairy Association.
ABOVE: *A broadcast from Oak Brook on WGN-TV. I'm with the 1966 Princess, Carol Ann Armacost of Maryland.* BELOW: *The 1968 American Dairy Princess pageant in Chicago. Florida's Elaine Marie Moore was the winner.*

WGN

ABOVE: *The National Pork Princess on "Top O' the Morning" on WGN-TV.*
BELOW: *Miss McHenry County, 1967.*

Courtesy: Don Peasley Photography

Courtesy: Don Peasley Photography

ABOVE: *What would a '60s beauty pageant be without the swimsuit competition?* BELOW: *Luci Baines Johnson was the guest of honor during her father's 1964 presidential campaign. This was at the Harvard Milk Days.*

Courtesy: Abernathy Photo Company

I've had the pleasure of interviewing dozens of Alices in Dairyland; from the '60s at left, to 1994 below and in the 2000s, bottom left.

All of the Alices are dairy farm girls and many of them competed against me in milking contests.

All those years of milking by hand when I was a youngster paid off: I was very hard to beat.

Courtesy: Lake Geneva Regional News

TOM WILKIE (center), director of sales at the Abbey, helped show the results of the cow milking contest conducted as part of the Abbey's Ice Cream Social activities. Also shown here are WGN Radio's Orion Samuelson (left), who won the contest, and Alice in Dairyland Barbara Ward, who came up short in the contest. The Ice Cream social was part of the Abbey's June Dairy Month celebration.--Regional News Photo

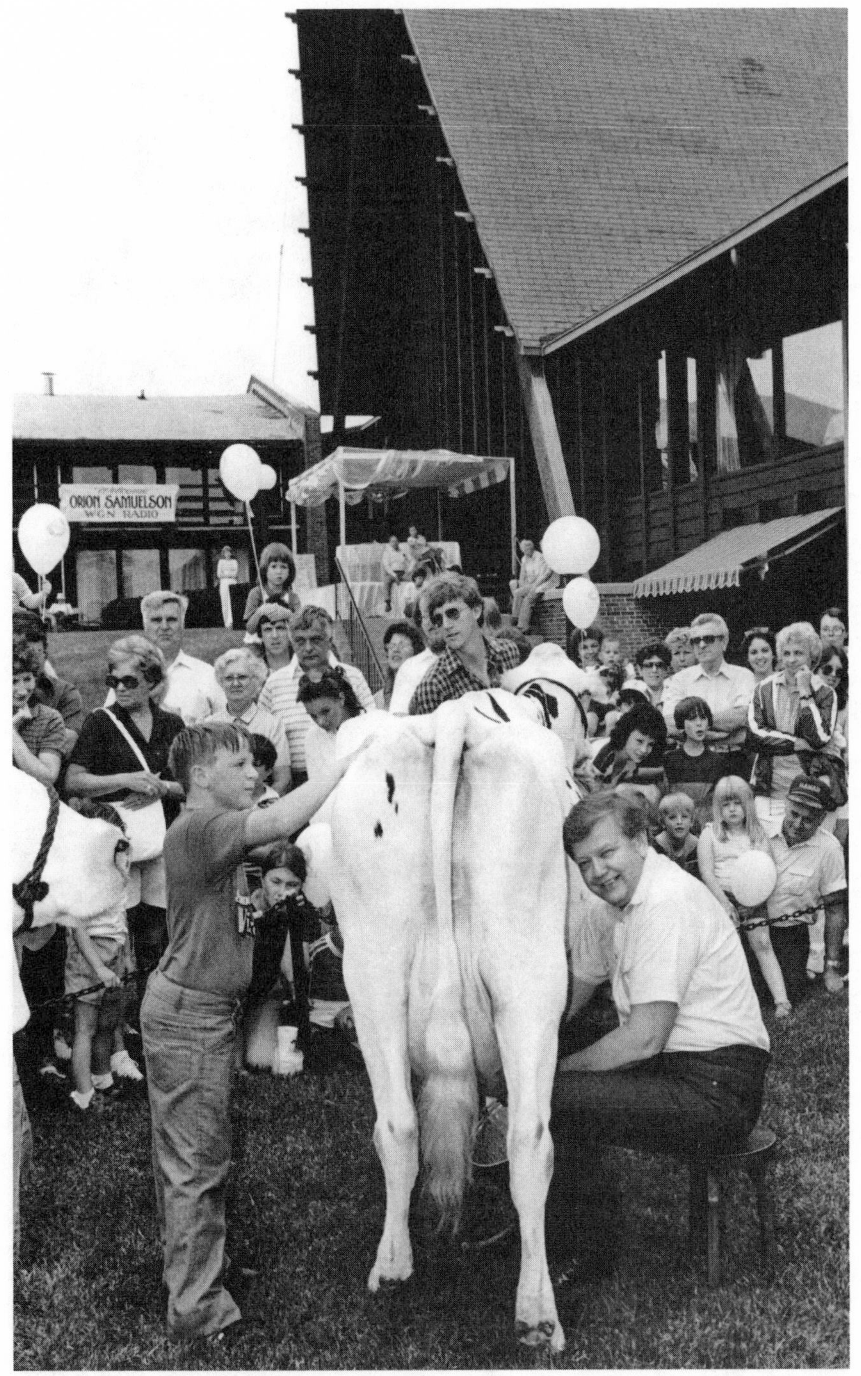

At The Abbey in Lake Geneva, Wisconsin.

Judging by my expression in the photo on the left, I was "udderly" delighted to have the opportunity to promote the dairy industry by demonstrating my milking prowess.

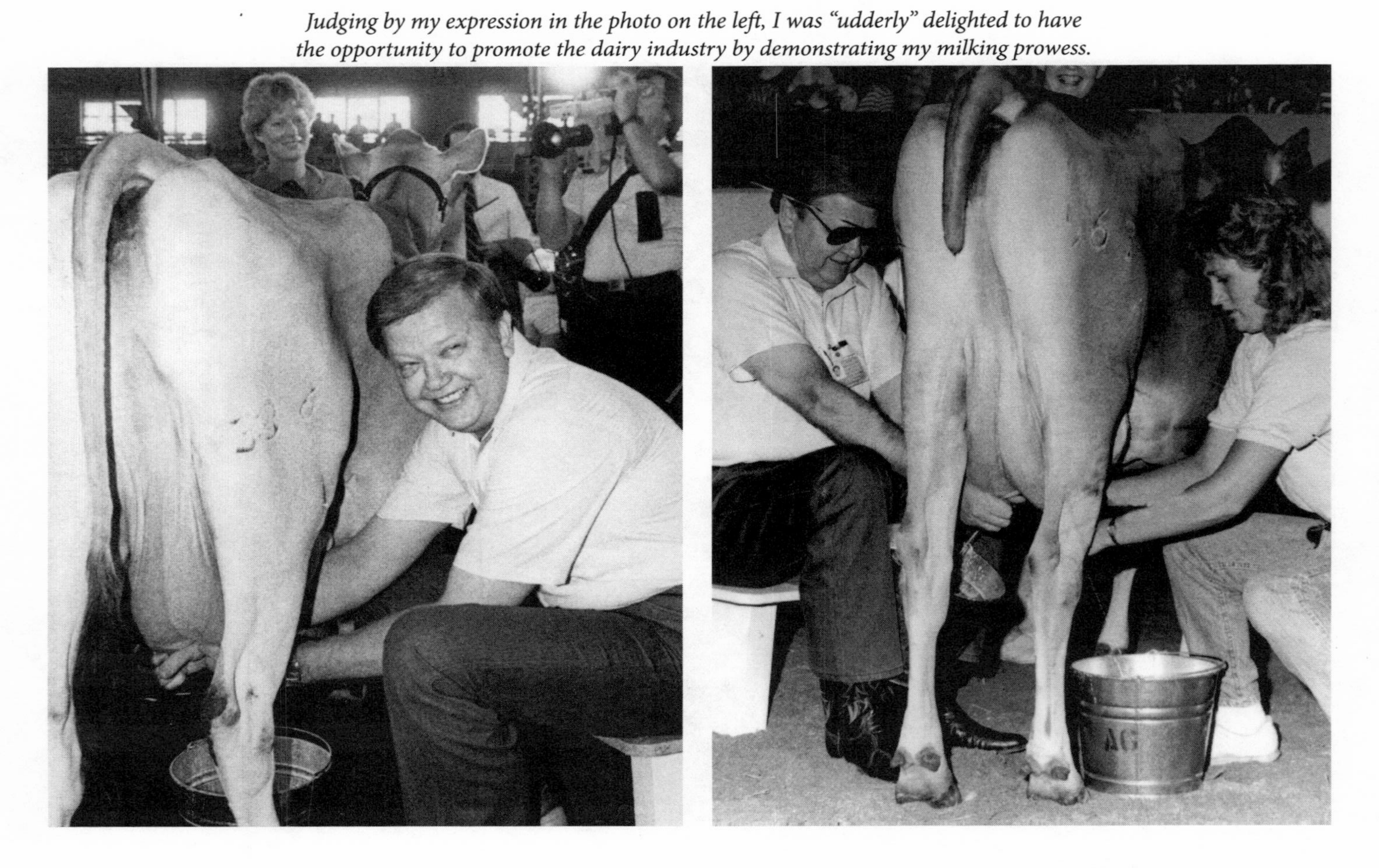

Ole and Lena were 90 years old and having trouble remembering things, so they decided to go to their doctor for a checkup. The doctor ran some tests and told them both they were just fine, but for their memory problems, they might want to start writing notes as reminders.

They thanked him and left.

The note strategy seemed to be working well. One night, while watching TV, Ole got up from his chair and headed for the kitchen.

Lena asked, "Vill you bring me a bowl of ice cream?"

"Ya, sure," Ole said, and began to walk away.

"Don't you tink you should write dat down?" Lena called after him.

"Naw, I can remember dat," Ole answered.

"Vell, I also vould like some strawberries on top. You'd better write dat down because you'll forget dat."

"I can remember dat," Ole said. "You vant a bowl of ice cream vit strawberries."

"Ya, but, I also vould like vipped cream on da top," Lena called after him. "I know you vill forget dat so you better write it down."

"I don't need to write dat down," Ole grumbled. "I can remember dat."

Twenty minutes later, Ole walked out from the kitchen and handed Lena a plate of bacon and eggs. She stared at the plate and snapped at Ole: "I told you to write it down. You forgot my toast!"

The restored F-20, loaded with Samuelsons.

Chapter Fourteen

The F-20

On a warm spring day in 1939, the tractor dealer pulled into the Samuelson farm barnyard with a delivery. He unloaded a bright red Farmall F-20. It was shiny and clean, without a speck of dust or grease, possibly the prettiest thing I'd ever seen in my five years of life. Because the country was preparing for World War II, there weren't any tires available, so the F-20 came with steel wheels. It wasn't until the late '40s before Dad was able to buy tires for it. When my legs got long enough and strong enough to push the clutch, Dad taught me how to drive it and from then on, that F-20 and I spent a lot of time together.

Dad and Mom paid $720 for the F-20. When they held their farm auction in 1964, the F-20 sold for $750. That was the last time I saw it, until Max Armstrong, my sister Norma and other family members, along with many other people staged an incredible surprise. I'll let Max pick up the story from here.

The story of Orion's F-20 starts with my Farmall Super H. In 1995, my mom and dad decided to sell off the rest of the equipment they had on the farm. I had a sale bill with me in Master Control at WGN Radio and a couple of the older engineers saw that old Farmall listed on it and said, "Kid, you can't let that get away." So, at my mom and dad's auction, I bid on the Farmall and I was the successful bidder.

The restoration of the Super H led to a tractor segment on our television

The carcass.

show, "This Week in AgriBusiness." "Max's Tractor Shed" is what we call it and that got Orion's sister, Norma, thinking about the tractor that their dad had. She sent me some pictures of it, and they weren't pretty! It had been sitting in their cousin's field in Wisconsin for at least 35 years and it was just a pile of junk, no other way to describe it. Everything was broken or missing. The steering wheel was broken. Parts just weren't there. Cattle had rubbed up against it. Birds had

David, my granddaughter Grace, grandson Matthew and a photo of my dad, Sidney, "Sam" Samuelson.

With Jim Irwin who retired as vice-president of Case-IH after 40 years with the company.

pooped all over it. It was just a skeleton sitting out there, which, of course, got me excited. I knew that even that old pile of junk had some importance to it. There isn't a farmer alive who doesn't have memories attached to the tractors they spent countless hours on as young men and women. Since it was Orion's dad's tractor, I decided to see what we could do with it. I sent a couple of guys up there, the Eipers boys, Larry and his nephew, Ben. They drove their flatbed truck to Wisconsin, hooked onto that old thing, winched it onto the truck and hauled it back down to Illinois. Some time later we were on a tractor ride, the Heritage Tractor Adventure in Morris, Illinois, and we had an auction to raise money for the Ag In The Classroom program. It helps tell the story of agriculture in classrooms all over the state of Illinois, including in city schools. We decided to see if there was any interest in that old carcass, and there sure was. The bidding became quite spirited that evening and the winning bid was almost $7,000! The new owners were members of the International Harvester Collectors' Club, Chapter 10 in Central Illinois. They said, "Yep, we're going to restore it," and it was a breathtaking restoration when you compare what it looked like before to what it looks like now. They spared no expense: the race car red paint alone cost $1,400, plus the labor. It is a magnificent, wonderful finish on that tractor; a far better finish than it would have had coming out of *the old Farmall plant.*

After the tractor had been restored, we had an unveiling for Orion. His wife, Gloria, son David and his family, and daughter Kathryn were all there and it was a wonderful family event with hundreds of others in attendance. When Orion saw that shiny, better than new F-20, tears streamed down his face, an affirmation of how much tractors mean to farmers and their families.

Ole and Lena's brother Sven are in the woods deer hunting. Ole bags a buck. After they dress the deer, they grab it by its hind legs and start dragging it through the trees back to the car. A game warden happens on the pair and, after checking their tags and admiring the buck, tells them that they are dragging the deer out all wrong. He told them that by dragging it by the rear legs, the snow, leaves and dirt were getting caught by the animals fur, and the horns are getting all tangled in the brush.

The warden suggests that they drag it by the front legs. Ole and Sven agree to try it and after a half-hour of this, Sven turns to Ole and says, "Boy dat game warden was right, Ole, it sure is easier dragging da deer dis way, but ya know, we are getting farder away from de car."

Air Orion

Chapter Fifteen

Air Orion

One of the problems with a single-engine airplane is that it's a single-engine airplane. Around midnight on August 22, 1996, Air Orion was cruising home from Grand Forks, North Dakota, where I had given a speech at a potato growers event. Pilot Jerry Lagerloef, cameraman Angelo Lazarra and I were almost home, flying over McHenry County, when the engine blew up. That began the most interesting seven minutes of my life.

First, though, let me describe my love affair with airplanes. It started when I was eight years old at the Vernon County Fair in Viroqua, Wisconsin. Dad paid a Piper Cub pilot $10 to take us for a ride over the fairgrounds. I was hooked, and my fascination with those magnificent flying machines grew steadily. Beyond the personal thrill I still get each time I become airborne, I could not have done my job nearly as well all these years without airplanes transporting me quickly and safely to events near and far.

American agriculture has benefitted from aviation, too. Most obviously, the crop dusters; daredevil pilots who skim the fields applying various products to increase yields. We depend on airplanes to deliver perishable products quickly to buyers around the world. With the foreign agricultural market becoming more important every year, producers log many miles on airplanes to conduct trade missions and increase sales of our farm products by personally meeting buyers and consumers in far away countries. And the combination of agriculture and aviation have been of great benefit to the U.S. economy. The United States racks up a huge negative trade balance every year, yet there are two areas where we consistently have a positive trade balance: airplanes and agriculture.

The beginning of Air Orion started late one night in 1979 in Pontiac, Illinois. It wasn't unusual, with my 40 to 45 speaking engagements a year, to drive to an event anywhere from 100 to 200 miles from Chicago, deliver my speech, then drive back to Chicago and get a few hours of sleep before going to work at WGN.

One night, after I spoke in Springfield, Illinois and was driving back by myself on old Route 66, my car was approaching the city limits of Pontiac. I woke up just in time to swerve to the right and avoid smashing under the trailer of a semi that was stopped at a red light. Yes, my eyes had closed. I had gone to sleep. By the time I stopped the car, I was on the shoulder, just about even with the tractor cab of that semi.

That really was a "wake-up call" that got me thinking, and I realized that given my ability to take a nap while driving, I was going to have to find a different way to get to my speaking engagements. At first, I hired a friend to drive me. Ben Hartmann and his wife Bonnie were dear friends from church and Ben loved to drive, so he volunteered to take me whenever

Air Orion pilot Jerry Lagerloef, his father Hans and pilot Phill Wolfe with me outside a hangar at the Scottsdale, Arizona airport.

he could to speaking engagements. Then, I was having a conversation with Paul Wallem, my lifelong friend and as close to a brother as I have, who was then the owner of an International Harvester dealership in Belvidere, Illinois. Paul was a pilot and had a Cessna 172. I mentioned my problem of getting to speeches. Paul said, "You know, Orion, I've been wanting to trade in the 172 and step up to a bigger and faster airplane. Would you be interested in going 50-50 on it with me?" I told him that I thought I could come up with the money and that I would like to do that.

In 1984, Paul found a four-passenger, turbocharged Cessna 210 in Texas. He flew it home and that was the beginning of Air Orion. It was ideal for me because not only would I use it for speaking engagements, it was big enough and fast enough that we could use it to fly a crew to television shoots.

An important decision I made when I bought Air Orion with Paul was that I would not get a pilot's license, because if I had, I knew that I would have taken chances that I should not have taken. If I had a speaking commitment in Indianapolis and there was a line of thunderstorms between Chicago and Indianapolis, I would try to find a way to pick my way through those thunderstorms so I could get there. That's the kind of decision that kills a lot of pilots. Always having a professional pilot in the left seat was the best decision I ever made because my mind goes elsewhere. I'm thinking of what I'm going to talk about and what kind of presentation I'll make. You can't be distracted like that when you're flying an airplane.

Paul Wallem flew me on several occasions, but my first regular pilot was Bob Harshbarger, a retired Air Force Major who had flown jet fighters in Vietnam. When I was on my second trip with Bob, he said, "Orion, if I have a heart attack up here, you need to know how to get me to a hospital." So on each trip, Bob taught me a little more about the different aspects of flying and, as a result, I know how to "drive" a plane. Over the years I have driven the plane a great deal, but I've always had a professional pilot in the left seat, literally a life-saving decision, as you'll be reading shortly.

Air Orion is great for our TV shoots, because it seats the pilot and me in front, two cameramen in the back seats and enough room behind them for two full sets of TV gear. We could drop into a short, 3,000 foot strip and have even landed in farm fields a couple of times, which wasn't a good thing to do because of the risk of damage to the prop. But having Air Orion, number one, saved me from killing myself by falling asleep behind the wheel of my car and secondly, saved me a lot of time and allowed me to spend a

lot more nights at home in my own bed because a location that would take me eight hours to drive, such as the Lake of the Ozarks in Missouri where I've had several speaking engagements, was a two-hour ride in Air Orion.

My second Air Orion pilot — well, third counting Paul Wallem — was Mike Hudgins and for the last 15 years, I've had two pilots: Jerry Lagerloef and Phill Wolfe. Between the two of them, they've been able to handle all of my needs to travel the country. I give all of my pilots one rule: there is no place I absolutely have to be. "Get home-itis" kills too many private pilots and I didn't want any part of that.

From my window in Air Orion, I've had the joy of watching the landscape go from the gleaming white of a snowy winter, to the colorless hues in early spring, to lush green in the heat of summer and to fields of gold at harvest time. I'll generally only use Air Orion if it's a short hop across the Midwest. If it's a longer flight, I'll fly commercial and I've logged over two million miles in the mostly-friendly skies.

On that night of August 22, 1996, we were cruising back to Illinois from North Dakota at about 11,000 feet with a tail wind, but when we got to Madison, the weather started to close in a little bit. The ceiling dropped to about 6,000 feet with some rain and mist, so Jerry guided the plane to 5,500 feet to get under the cloud cover. We were eight minutes from landing at the airport in Wheeling, Illinois, at that time known as Palwaukee Airport.

GALT FLYING SERVICE, INC.
5112 GREENWOOD ROAD – GALT AIRPORT
WONDER LAKE, ILLINOIS 60097 • 815-648-2433

AERO COMMANDER®
CHARTER – RENTAL
FAA APPROVED AND VETERAN APPROVED
FLIGHT SCHOOLS

Cessna

FAA APPROVED REPAIR
STATION 3102

October 17, 1996

To: Orion Samuelson

Sir:

Enclosed you will find the remaining bill for the repairs to your T210 Cessna. The bill shows an engine core charge of $3500.00. In a telephone conversation with Continental Motors, a factory representative stated the full core charge of $7000.00 is never billed if the engine is returned regardless of condition. He stated that the usual charge for damage such as yours is $3500.00.

In addition to the engine, your aircraft [illegible] flushed and the governor was replaced [illegible] contamination remained. The [illegible] and did not require repl[illegible] beyond service li[illegible] as well as [illegible]

[illegible] connecting

[illegible] please call at the above [illegible] and the staff at GFS thank you for [illegible] appreciate the chance to help you. Hope [illegible] sometime in the future!

Respectfully,

Bob Russell
Manager/IA

Pd 11-8-96
#1381

"We at GFS wish to extend our sincerest admiration to you and Mr. Lagerloef on the good job of landing the aircraft after such a catastrophic situation."

Angelo was asleep in the backseat and I turned to Jerry and said, "Boy, this has been a good night for flying. There's been no turbulence and we've had a nice tail wind... " And just then, from the front of the airplane, came a loud noise, a mini-explosion. The airplane started shaking violently, like it was going to come apart. The prop was still spinning, but I glanced at the gauges and saw that our power had dropped to about 10 percent. I looked at Jerry and said, "What in the hell was that?" He said, "It's the last thing you want to hear at midnight at 5500 feet above the ground in northern Illinois." Jerry immediately got on the radio and declared, "Mayday" and got instant responses from the Rockford airport, O'Hare field and also from Air Traffic Control located in Aurora. After Jerry gave our exact location, Air Traffic Control said it would take jurisdiction.

We were losing altitude at the rate of about 400 feet per minute and Angelo was wide awake by then, looking about as white as any person could. We're all looking at the instrument on the dash that, at the touch of a button, would give us a list of the ten closest airports and what heading to fly to get there. The first one that came up was the McHenry Farms Airport. I told Jerry, "That won't work, it's a sod strip with trees on both sides and no runway lights." Air Traffic Control radioed that we had Lake in the Hills Airport eight miles to the east. Jerry replied, "This airplane won't stay in the air for eight miles. We need to find something closer."

Meanwhile, Jerry is telling me, "Look for water, Orion, we'll put it down in water." I said, "Jerry, I can't swim!" He said, "Don't worry, I'll save you!"

It's pitch black out my window except for an occasional yard light indicating a farm. But then, off in a distance about three miles away, I see runway lights! At about the same time, the instrument panel and Air Traffic Control said Galt Airport was close by. Maybe not close enough. We're still losing altitude and because of the wind direction, Jerry had to turn the plane around to make an approach. The plane continued to shake violently, but the prop was still turning and Jerry managed to get it lined up with the runway and then radioed Air Traffic Control what our latitude and longitude were and said, "If we don't make it, this is where you'll look for us." I thought, okay...

We made it, but barely. Just after the plane touched down, the propeller stopped spinning and the engine froze. The plane rolled to a stop. The three of us got out and went our separate directions to say, "Thank you, God, for putting us on the ground." God decided he wasn't done with me yet, I guess.

After I called the McHenry County Sheriff's office, the owner of the airport, Art Galt, was called. He came out of his farmhouse, wearing a nightshirt, night cap and tennis shoes. As he walked around the corner of the hangar in the misty light from the one hangar light that was on, he said, "Samuelson, what in the hell are you doing at my airport at midnight?" I said, "Art, thank you for leaving the runway lights on. It saved our lives." Art was a dairy farmer who flew and built a very nice 3,000-foot asphalt runway next to a cornfield. About ten years earlier, he nearly lost a plane one night and decided the lights would stay on after that.

An inspection of the engine showed that a piston had come through the side of the block and we lost all our oil. $42,000 later, we had a new engine and Air Orion still flies today and does just a beautiful job. I'm blessed with some excellent pilots.

A footnote to the story: the pilots always send me their bills after our flights. When Jerry sent his, he added $5 for a new pair of Jockey underwear.

Ole and Lena had vacationed in Norway and were ready to fly home to the States. Shortly after their four-engine airplane took off, the captain announced over the PA system that there had been a problem in engine number four and he had to shut it down, but, no worries, they still had three good engines. They'd just be a little late getting home. Ole told Lena, "Ya, dat's okay. Ve yust get a little longer vacation."

An hour later, the captain came on the PA system again, announcing that there were now problems with engine number one, but again, not to worry, they'd just be arriving about an hour later than scheduled. Ole saw that Lena had a concerned look on her face, so he gave her hand a reassuring squeeze. "Ve'll be okay."

The long flight continued without problems and Ole and Lena drifted off to sleep. They were jolted awake by the captain announcing that there had now been problems with engine three, but, again, he said the airplane was designed to fly on one engine and everything would be fine, except that their arrival time was pushed back by two more hours. Ole, clearly annoyed by this latest news, looked over at Lena, and said, "For cryin' out loud. If another engine goes out ve'll be up here all night!"

PASS ON THE GIFT
HEIFER®
INTERNATIONAL

FRB
Foods Resource Bank
A Christian Response to World Hunger

Wreaths Across
America™

Chapter Sixteen

Thanks and Giving

One holiday tradition I started decades ago was reminding my radio, television and newspaper audiences each year around Thanksgiving to be grateful for America's farmers and ranchers. The men and women who put food on our table are so good at what they do, it's easy to take them for granted. And it isn't just food. They grow fiber for the clothing on our backs, wood for the roofs over our heads and ethanol for our gas tanks.

Our gratefulness extends beyond the agriculture sector, though. From the pioneers who endured harrowing hardships to settle this great country, to the pioneers of industry, science and medicine who have made our lives richer and more comfortable in countless ways, we owe them all not only our gratitude, but our lives.

In 2007, I was in Washington, DC a few days before Christmas and some friends at the Department of Agriculture told me that if I ever had the chance, I should visit Arlington National Cemetery around Christmas because it's so beautiful. I didn't have time on that trip, so when I returned home, I searched the Internet and found photographs of Arlington that were stunning. Rows and rows of white stone markers in the snow-covered cemetery were each decorated with magnificent green wreaths and big, bright red bows. Always the curious one, I wanted to find out where the wreaths came from, who paid for them and, as my friend Paul Harvey famously said, "the rest of the story."

The "Wreaths Across America" story started in the early '90s when Worcester Wreath Company, a for-profit business located in Harrington, Maine, began a tradition of placing wreaths on the headstones of our nation's fallen heroes at Arlington. In a phone conversation with the owner, Morrill Worcester, he told me that when he was 12, he was a paperboy for the *Bangor Daily News* and won a trip to the Nation's Capital where he visited Arlington Cemetery and it made a lasting impression. Later in life, he realized that his success in business was due, in part, to the sacrifices made by our military veterans and was always looking for a way to say "thank you."

In 1992, when he had a surplus of unsold wreaths, he loaded 5,000 of them on a truck, drove from Maine to Arlington Cemetery and began placing them on the stone markers in an older section that didn't see as many visitors. Cemetery officials were skeptical at first and were concerned about who would pick them up and dispose of them when the Christmas season ended, but Morrill said when they saw the beauty of his work — lush, green wreaths accented with the red ribbons — they thanked him and said they would take care of the clean-up.

In the years since, more than 100,000 wreaths were placed on the Arlington grave markers by volunteers. Thousands more were donated to the Maine Veterans' Memorial Cemetery and to state and national veteran cemeteries across the country. The word spread and demand became so great that in 2007, "Wreaths Across America" was formed and there are now over

Courtesy: Wreaths Across America

700 cemeteries and monuments receiving wreaths each Christmas season.

It is a not-for-profit foundation whose mission is: Remember, Honor, Teach. Remember the fallen. Honor those who serve. Teach our children the value of freedom. The story and photos can be found at www.wreaths-across-america.org, where you can also donate to help what Morrill Worcester started live on.

There are many other non-profits who need and deserve your help. I'll mention a couple which are among my favorites because they are agriculture-based and their focus is helping people learn how to help themselves.

First, Heifer International. This is a program that started in 1944, at the end of World War II by Dan West, an Indiana farmer who had served as a relief worker in Spain during the Spanish Civil War in the late '30s. He became frustrated because of the very limited rations of milk that were available to give to refugees. When he returned to the United States, he started Heifers For Relief, an organization dedicated to providing permanent freedom from hunger by giving families livestock and training so that they "could be spared the indignity of depending on others to feed their children." His philosophy was based on the proverb, "Give a man a fish; you have fed him for today. Teach a man to fish; you have fed him for a lifetime."

Dan West's "Give not a cup, but a cow," set the example for what would become Heifer International's model for sustainability. Each family receiving a heifer or female animal agreed to study animal husbandry and also agreed to donate a female animal offspring to another family, which would then do the same. Dan's vision was that one animal's impact would go far beyond the original investment. The first shipment of 17 "Heifers For Relief" went from York, Pennsylvania to Puerto Rico in 1944.

Since then, Heifer has helped more than seven million families in 125 countries, and that includes the United States. Each year, Heifer assembles what it calls the "The Most Important Gift Catalog in the World," where you can buy an animal or some other agricultural commodity that will go to a person in a developing country or even a poverty-stricken rural area in this country to help them produce food and income. Besides giving a heifer, you will find in the catalog the opportunity to give goats, sheep, llamas, rabbits, ducks, chicks, pigs, tree seedlings, honeybees, water buffalo, camels and more. The cost is as little as $20 for a flock of chicks to as much as $850 for a camel. Or, if you're feeling like Noah, you can give the Gift Ark, 15 pairs of animals for $5,000.

The reasons I strongly support the Heifer mission include that it gives people dignity and the ability to feed themselves instead of just accepting a handout. It also is the ideal gift for the person on your Christmas list who has everything. Gloria and I have gifted several of our friends over the years, putting their names on gifts of hope for people they will never know. It's also an easy way to shop. You can get full information, including the catalog, at the Heifer International website, www.heifer.org. When Heifer celebrated its 50th anniversary in 1994, I was humbled to be named its "Man of the Year," an honor of which I'm especially proud because of my admiration for the great work that Heifer does in using agriculture to help others help themselves.

Another organization in which I am involved is the Foods Resource Bank. It, too, helps people in developing countries feed themselves, but in a different way. In this country, rural churches and city churches become partners, and in the spring fund the planting of 30, 40 or 50 acres of grain. Rural church farmers donate the acres and do the work; city church members provide money for the seed, fertilizer, fuel, etc. At harvest time, all the church members come together at the farms, have church services and lunches, then harvest the crops, which are then sold.

The proceeds are used in different ways. In the case of my church, the money went to the digging of wells to provide clean water for drinking and irrigation in a rural village in India, again helping people feed themselves. There is another benefit to this program; it brings city people to a farm at harvest time so they can see firsthand what farmers do to put food on their table and get to know them personally. And kids and adults love riding in the combine cabs! The website: www.foodsresourcebank.org.

And one more non-profit doing worthy work is AABB, the All-American Beef Battalion. It was established in 2007, and its goal is to support our troops fighting the war on terror and also promote American beef. They accomplish the task by organizing and sponsoring steak feeds and entertainment programs for service members and their families. Since 2007, the organization has served more than 100,000 steaks to our troops. AABB volunteers roll onto military bases with several trailer-mounted grills and refrigeration trucks. One of their busiest stops each year is Fort Bliss, Texas, where they grill over 5,000 steaks for the troops and their families. That effort is still going strong. More information is available on AABB's website, www.steaksfortroops.com.

One more Thanksgiving note: when you're gathered around the holiday table with your family and friends, giving a prayer of thanks and acknowledging the hands that prepared the food, would you also please give thanks for the hands that produced the food? Thank you.

Ole had been rushed to the emergency room, and after examining him, a doctor took Lena aside and said, "I don't like the looks of your husband at all."

Lena said, "Oh ya sure. Needer do I, Doc, but Ole knows how to farm and is goot wit da kids."

Photo: Trond Isaksen / Nobel Peace Center, Oslo

Nobel laureate Norman Borlaug at the Nobel Peace Center in 2006.

Chapter Seventeen

Norman Borlaug and Harold Brock

Over the four decades in which I was privileged to know Dr. Norman Borlaug, we shared a few dozen conversations in radio studios and in front of television cameras. I have often said that we had three things in common: we are both of Norwegian ancestry, both grew up on farms without electricity and both went to one-room country schools. Believe me, that's where the similarities came to a screeching halt. He did the hard work that changed the world and I just talked about it.

Fascinating doesn't begin to describe this man. He was awarded the Nobel Peace Prize in 1970 for developing varieties of wheat and rice that would grow in arid climates as well as the tropics. Because of his research work in Mexico and India, he took world grain production from 692 million tons in 1960 to 1.9 billion tons in 1992. Most impressively, that increase in world grain production came with using only one percent more land because of Dr. Borlaug's accomplishments.

Yet, Dr. Borlaug was a humble individual and wouldn't talk much about the countless honors and awards he received. He is one of only five people in history to win the Nobel Peace Prize, the Presidential Medal of Freedom and the Congressional Gold Medal. *TIME* magazine in 1999, named him one of the "100 Most Influential Minds in the 20th Century." His was a mind that did not stop working. Up until a few weeks before his death, he was continuing his research at Texas A & M University. The United Nations officially credited him with saving a billion people from starvation. The title

Norman Borlaug in 1970, when he won the Nobel Peace Prize.

of his biography published in 2007 (a must-read) says it all: *The Man Who Fed the World.*

From his humble beginning on an Iowa farm, Dr. Borlaug made an impact that will never die. He was honored at a White House dinner after he

Harold Brock at Union Grove, Wisconsin in 2009.

received the Nobel Peace Prize in 1970. I was an invited guest at that dinner and it remains one of the highlights of my life.

Several years ago, I visited the university in India where he did much of his work in the '60s. At the entrance to the building was inscribed one of his famous sayings: "Everything else can wait, agriculture can't."

Dr. Norman Borlaug, the Father of the Green Revolution, a giant of a man and one of my personal heroes, passed away in September, 2009 at the age of 95.

Another of my personal heroes was Harold Brock, who, as much as any person, was the father of the modern tractor. I had the opportunity to interview him twice and consider him to be one of the most fascinating people I've ever met. In 1929, at the age of 15, Mr. Brock enrolled at the Ford Trade and Apprentice School in Dearborn, Michigan where he studied to become an engineer. Each student was assigned to an apprentice foreman, or mentor, and his was none other than Henry Ford. And Ford's mentor was Thomas Edison. Mr. Brock said that during his early years at the school and later at Ford, he had the opportunity to be a fly on the wall as Ford learned from Edison. Mr. Brock told spellbinding stories about that time in his life, including his interactions with many of the great minds of the early 20th century with whom Ford consulted, such as George Washington Carver, Harvey Firestone, Luther Burbank and a British inventor, Harry Ferguson.

Carver was a big influence on Henry Ford. Ford was a fan of the soybean, which was mostly a cover crop back then that was plowed under, and its many possible uses. His goal was to grow all the materials needed to build a vehicle. Mr. Brock said that Ford assigned him to design a tractor seat made out of soybeans, which he did. But there was a problem. "Orion, I had to tell Mr. Ford that his soybean seats were getting chewed up by the rats and mice. That ended that." Much cheaper plastics also ended, or at least delayed, Ford's soybean research.

As for the tractor, Mr. Brock said Henry Ford wanted to replace animal power, but Ford knew that he had to build a replacement farmers could afford. His order to Mr. Brock, who was appointed chief of tractor engineering, was to design and produce a tractor that wouldn't sell for any more than the farmer would pay for a team of horses, including the cost of harnesses and feed. Within six months, Mr. Brock's design rolled out of the Ford tractor plant at a price of just over $500. His N Series Ford tractors were the first to feature a three-point hitch, Harry Ferguson's idea, and a rear

PTO (power take off) shaft.

When World War II came around, Mr. Brock's tractor designs were put on hold. He was designing war vehicles instead. When the war ended, Mr. Brock finished work on what would become one of the biggest selling tractors of that time, the little red and gray 8N. He said the red chassis was to keep rust from showing and the light color for the sheet metal was to keep droppings from chickens that would surely roost on the tractors from being so noticeable. What sealed the decision for Mr. Brock was when he saw the red and gray combination on one of his wife's dresses.

Harold Brock remembers Ford as a good boss, but a man who wasn't worried too much about money. When Mr. Brock was making 13 cents an hour, he asked Ford for raise. Ford told him, "Money's not the objective of life. The important thing is, do you like your job?" Mr. Brock told him he loved it. Ford said, "Then don't worry about the money." Mr. Brock thought that was funny, coming from a millionaire. He worked at Ford for nearly 30 years. In 1959, 12 years after Henry Ford died, Mr. Brock went to work for John Deere, developing tractors at its Waterloo factory, including the iconic 4020, another big selling tractor. Mr. Brock became Deere's first Worldwide Director of Engineering and was instrumental in Deere overtaking International Harvester as the industry leader.

Along with his contributions to the farming world, Mr. Brock also had a significant impact on young people and their education. In 1965, Harold Brock helped bring a two-year technical college to northeast Iowa, the Hawkeye Technical Institute in Waterloo, which is now Hawkeye Community College. It now has nearly 6,000 students and Mr. Brock was active for more than 30 years in fundraising and other support for the college. He also helped start Junior Achievement in Iowa.

Mr. Brock died in 2011 at the age of 96, but right up until the end, he was on the move, telling stories of those amazing days at Ford and Deere.

Ole was the pastor of the local Norwegian Lutheran Church, and Sven was the minister of the Swedish Covenant Church across the road. One day they were seen pounding a sign into the ground, that said:

DA END ISS NEAR! TURN YERSELF AROUND NOW BEFORE IT'S TOO LATE!

Just then, a car sped past them and the driver leaned out his window, shaking his fist and yelling, "Leave us alone, you religious nuts!"

*Moments later, Reverend Ole and Pastor Sven heard screeching tires and a big splash. Reverend Ole turned to Pastor Sven and asked, "Do ya tink maybe da sign should yust say '*BRIDGE OUT*?'"*

With my longtime friend, Paul Wallem, in 2002 at the University of Illinois where I was awarded an honorary doctorate degree.

Chapter Eighteen

Finally, a College Degree

Only 50 years passed from the time I entered the University of Wisconsin until I received my degree from the University of Illinois. One reason it took me so long is because I didn't work for it, at least in the traditional sense. As I mentioned in an earlier chapter, I lasted only a few weeks at UW-Madison in 1951 before deciding that I wasn't going to be taught the broadcasting skills that I wanted to learn. So I remained three years and nine months short of a college degree until the age of 67 when the University of Illinois conferred upon me the honorary degree of Doctor of Humane Letters. It was truly one of the special days of my life to stand on the stage in Assembly Hall at the University of Illinois and receive my first college degree.

The University doesn't award such degrees with a great deal of frequency and, of course, I questioned what I had done to earn it. While I had been involved for decades in promoting agriculture and the great work done at land-grant universities like the University of Illinois and its College of Agriculture, now known as the College of A.C.E.S., which stands for Agriculture, Consumer and Environmental Sciences, I was just doing my job. I do owe a debt of gratitude to David Chicoine, who was most instrumental in doing the paperwork and legwork that took my name to the honorary degree committee. David was the Dean of the College of A.C.E.S. at that time and has since moved on to become president of South Dakota State University.

As with most awards I have received, I felt that I had to justify why I received the degree, so I've become deeply involved with the U of I, helping with fundraising efforts and highlighting the many activities and research programs there.

Adding to my connection to the University, in honor of my 40th anniversary at WGN, two good friends of mine, Chuck Bloomberg and John Huston, started a college scholarship fund in my name at the University of Illinois. It has grown over the years to a level where we can present three scholarships annually to students in the College of A.C.E.S. The recipients are chosen by the university, I have nothing to do with that process. It is now named the Orion and Gloria Samuelson Scholarship Fund and our involvement has been a real thrill for us because we firmly believe, just as I was helped as a youngster by my Vo-Ag teacher and others, it's important to give back. The ability to help via the scholarship fund is humbling and gratifying.

In 2002, good friends Chuck Bloomberg (left) and John Huston worked with others to establish a scholarship fund in my name at the University of Illinois.

On a hot, muggy summer day, Lena came home to find Ole on a ladder painting the outside of the house. He was sweating like crazy.

She yelled up at him, "Ole, vaht you doin' up dere dressed like dat?"

Ole shouted back, "Right dere on da can it says, 'Put on two coats.'"

CHANNEL EARTH

Chapter Nineteen

A Bump in the Road

On Easter Sunday, 1996, I was sitting in the backyard of a friend's home in Scottsdale, Arizona and made the decision to start *Channel Earth.* My host, Gene Koch, who had a lot of experience in business, both domestically and internationally, said, "Let's do it. I'll do the business plan."

Several months before that, I'd noticed the increasing number of specialty channels popping up on cable television. There was the Weather Channel, the Food Network, hunting and fishing channels and so on. I started thinking to myself that the most important basic industry in this country is agriculture so why shouldn't there be a channel devoted to agriculture?

I had discussed the idea with many friends before that Easter Sunday meeting which was the beginning of what became *Channel Earth.* We spent a lot of time working on the concept, but also on the physical development. We rented a full floor in a building about a block away from the Tribune Tower and the WGN Radio studios. By 1997, the Tribune and WGN no longer were involved in producing and distributing the weekly television show Max and I had done for most of two decades, *U.S. Farm Report.* It was being produced by *Farm Journal Magazine*, whose management wanted Max and me to move to South Bend, Indiana to host the show. We said no, and would eventually start another weekly show, *This Week in AgriBusiness,* so WGN's management agreed that if I had the time, I could continue to do the market reports on WGN Radio while developing the agricultural television channel.

Our concept included setting up a camera on the trading floor at the Chicago Board of Trade during trading hours and cover both the CBOT and the Chicago Mercantile Exchange. We planned to do extensive agricultural weather forecasting for the nation. And we would do lengthy interviews with people who normally get just a 15-second sound bite on television. For example, I did a 30-minute program called *Samuelson's Journal*, interviewing people involved in agriculture.

We had a great staff. Of course, Max Armstrong was in on it from the beginning, as was Lyle Dean, who helped us with programming and production. We recruited some other talent that we would use as anchors and reporters. And we found several investors willing to put in money, so we decided to go forward.

We worked with the National Rural Electric Cooperative Association, which had a separate telecommunications group working on bringing the Internet to rural communities. It helped us get a DirecTV satellite channel that we would use 12 hours a day and the rest of the time would be used for other programming.

We built a beautiful studio, we signed up some sponsors and at 5 a.m. on March 28, 1997, television's first news and information channel devoted exclusively to serving the unique needs of farmers, ranchers and all of rural America signed on. One of my quotes to the media covering our launch was: "We're excited about the opportunity to meet the information needs of people in rural America. The satellite technology available today gives us the best way to reach an audience beyond the city limits that deserves a steady, reliable, up-to-the-minute source of critical information on issues that affect their lives every working day. And we'll provide some fun, too."

It *was* fun, for a while. All of the pieces seemed to be in place for a successful operation. We were getting excellent response from farmers and ranchers across the country who liked the idea of a TV channel that was devoted to their needs.

We thought we were doing okay, but there were challenges, including one that was unexpected and costly. There was a problem getting television equipment delivered to our new studio in time for the launch. As a result, we had to rent for several months, at a high cost, two satellite trucks to get the programming on the air.

Selling the concept to advertisers was a challenge because it was a new concept, or because their budgets had already been committed elsewhere. But we did have some sponsors who came on board with us and things were going well.

Of course, we had taken out loans and our largest one was with

LaSalle Bank. About eight months after we started, LaSalle Bank decided that we weren't going to make it. The loan officer in charge of our account was the toughest businesswoman I'd ever encountered. She really stayed on top of us and turned up the heat. We were doing all we could, including offering special deals to advertisers if they would pay up front so we could satisfy LaSalle's demands. But finally, about 10 months after we signed on, the loan officer told us we had two days to make good. We couldn't.

On the third day, she walked into our TV studio and said, "Mr. Samuelson, you will go on camera and you will tell the viewers that this is the end of *Channel Earth*."

That was the toughest thing, from a business standpoint, that I've ever had to do. But I did it, she shut us down and that was the end of *Channel Earth*. There was no resurrecting it because investors had lost money, including me; I lost a great deal of money.

As I look back and ask what could we have done differently, one mistake was building a Cadillac when we should have built a Chevrolet. We built a fine channel with very capable people. We should have started much smaller and grown it.

A man I've known for at least three decades, Patrick Gottsch, was a Nebraska farm boy who became a trader at the Chicago Mercantile Exchange. In 1982, he moved back to the Cornhusker state and got into the satellite communications business. In the '90s Patrick tried to start a satellite TV agricultural channel and had failed twice. But the third time, he started RFD-TV, which signed on in 2000 and is very successful. The weekly show Max and I do, *This Week in AgriBusiness,* is carried by Patrick's channel. I give him a lot of credit for persevering, because after my one attempt, I said, no, that's it, because too much of my retirement money was eaten up by *Channel Earth*. I don't regret doing it because I would have always wondered, "What if... ?"

A lot of the people who worked for us those ten months tell me today that they are grateful for the experience, even though it ended prematurely and abruptly.

When I spoke to the 2012 graduating class at the University of Illinois in Champaign/Urbana, I suggested to the students that each of them, like me, had been writing a book since the day they were born. And one chapter we all had in common could be titled "Bumps In The Road." My biggest financial bump was named *Channel Earth.*

While we can and should dream big, not every dream will become a reality and as disappointing and painful as those bumps can be, it's important that we try, as pointed out in this Teddy Roosevelt quote, one of Gloria's favorites.

"It is not the critic who counts; not the man who points out how the strong man stumbles or where the doer of deeds could have done better. The credit belongs to the man who is actually in the arena, whose face is marred by dust and sweat and blood, who strives valiantly, who errs and comes up short again and again, because

there is no effort without error or shortcoming, but who knows the great enthusiasms, the great devotions, who spends himself for a worthy cause; who, at the best, knows, in the end, the triumph of high achievement, and who, at the worst, if he fails, at least he fails while daring greatly, so that his place shall never be with those cold and timid souls who knew neither victory nor defeat."

Teddy Roosevelt
"Citizenship in a Republic,"
Speech at the Sorbonne, Paris, April 23, 1910

Ole was staggering home after a night at the Norski Tavern. His Lutheran pastor, Reverend Ingvoldstad, happened to be driving by, stopped, and backed up. "Ole," he called out, "do ye vant a ride home?"

Ole accepted and when the pastor's car pulled up in front of his home, Ole asked Reverend Ingvoldstad to step inside for a minute.

"Oh, well, I suppose, Ole, but why?"

Ole explained, "I vant Lena to see who I have been out vit."

Since 1965, I've been privileged to emcee the annual National Outstanding Young Farmer banquets. Each year, the organization has presented me with an honorarium of some sort, for which I'm very grateful; perhaps a donation to my scholarship foundation, or, as shown above, a beautiful plaque.

Chapter Twenty

Assorted Memories

Organizations across the agricultural spectrum have provided me with the opportunity to be involved in telling their stories. I have attended conventions of just about every farm and commodity organization and interviewed their leaders and members. Some memories...

American Farm Bureau Federation - The first AFBF President I interviewed was Charles Shuman, a soft-spoken conservative Illinois farmer who commanded attention in Washington, DC and is the only agricultural leader I know who was on the cover of *TIME* magazine. The organization has been blessed with very strong leaders who followed Charlie.

National Grange - Herschel Newsom was its longtime leader and carried the title of Master of the National Grange. He was a challenging interview because he never ended his sentence with a period; it always ended on an "up" note, which made it impossible to edit cleanly for broadcast purposes.

National Farmers Union - The fiery leader I enjoyed interviewing was Jim Patton, always recognizable because of his black eye patch, and a staunch supporter of family farms.

National Farmers Organization - The rebellious organization, born in the late '50s, believed that to raise farm prices, farmers should dump milk and not sell livestock. The President was a Missouri farmer, Oren Lee Staley, who, in one of my interviews, I called the "Billy Graham of farm leaders" because of his ability to fire up a crowd. My friend, Don Muhm, the outstanding farm reporter at the *Des Moines Register,* wrote an interesting history of Staley and the NFO.

Farm Safety 4 Just Kids - Marilyn Adams, Iowa farm wife and mother, lost her 11-year-old son in a grain wagon accident in the '80s. Marilyn turned her grief into something positive by forming this organization to work with farmers and ranchers to make their places safer for children and spare them the grief of losing a child. She is truly an inspirational lady and I was proud to serve on her Board.

National FFA Alumni Association - I was involved in the Chicago meeting on a very cold winter day when this association was born. I joined other former FFA members who didn't want their FFA experience to come to an end, and felt former members could stay involved by supporting the teachers and students in their local FFA Chapter. Today, it is a very strong and active organization.

National Outstanding Young Farmers - This farmer recognition program started in the mid-'50s when an Iowa farmer, who was a member of the Jaycees, convinced his fellow Jaycees to start a national program to, guess what? Tell agriculture's story in a positive way; something we are still trying to do today. I have emceed the National Awards Banquet every year since 1965. It's one of the highlights of my year because now I have a family of friends across the nation who are some of the world's best farmers.

National FFA Convention and National 4-H Congress - I have attended all but two National FFA Conventions since 1958. I attended the National 4-H Congress in Chicago from 1957 until it ended, which was very sad for me because it was such a positive event for young 4-H members. I frequently run into people today who remember their experience at Congress and being interviewed by me. Meanwhile, I continue to host the RFD-TV gavel-to-gavel coverage of the National FFA Convention.

There are many other organization stories I could share, but I guess that will have to come in my next book. So, I end with 4-H and FFA, as I mentioned in Chapter 12, the two best leadership building organizations for young people on the planet today. I was privileged to belong to both as a young person on the Wisconsin dairy farm and because of that experience I have been blessed with the ability to dream big and fulfill those dreams.

Ole was seriously ill. He was slipping in and out of a coma. Lena stayed by his side and each time Ole regained consciousness, he saw her sweet, concerned face.

During one of his lucid moments, he motioned for her to come close so he could tell her something.

"Ya know, Lena, you have been wit me tru all da bad times. Ven I was fired, you were dere to support me. Ven my business failed, you were dere. Ven I got shot dat time, you were dere. Ven vee lost da house, you were dere. Ven my helt started failing, you were dere."

Ole paused for a moment to catch his breath. Tears were streaming from his eyes. "You know vaht, Lena?"

"Vaht, Ole?" she said gently, her eyes beginning to fill, too.

"I'm beginning to tink dat you're bad luck!"

Sincerely,
STATE OF ILLINOIS
OFFICE OF THE GOVERNOR
SPRINGFIELD 62706
My best to you always.
Sincerely,
JOHN R. BLOCK
Secretary
MIKE JOHANNS
Secretary of Agriculture
Paul Harvey News
The Ward L. Quaal Company
P.O. Box 336
Winnetka, Illinois 60093
Mr. Orion Samuelson
WGN Radio
435 North Michigan Avenue
Chicago, Illinois 60611
George W. Bush

Chapter Twenty-One

Letters

One day, I'm afraid letter writing will become a lost art; it's certainly dying. E-mail has nearly driven the U.S. Postal Service out of business. If you're over the age of 30, you may fondly remember the anticipation of running to the mailbox, the thrill of finding an envelope addressed to you and unfolding a handwritten, personal letter in your hands that someone painstakingly, perhaps lovingly, created for you. That has been replaced by a chime from your computer, telling you another e-mail has arrived and there's a good chance it's spam. Texting, tweeting and posting have reduced our written communication to phonetic mush.

The following letters were selected from many sent to me over the years by people who were important in my life. Some of them, like those written by Wally Phillips, are beautifully crafted works of art at which I marvel each time my fingers touch the embossed linen stationery. The note from Bob Collins, scratched on a WGN note pad, makes my heart ache each time I read it. I can still hear Paul Harvey's voice, pregnant pauses and all, when I re-read his carefully typed letters. I chuckle at the ones from Earl Butz and smile with appreciation for letters of congratulations and praise written by people like Ward Quaal, Presidents Reagan and both Bushes, and Secretaries of Agriculture Jack Block and Bob Bergland.

I hope you enjoy reading them.

Wally Phillips

Dear Orion:

The only thing wrong with our "Passing Parade" is that it does. And much too swiftly.

The good part is that along the way you get to accumulate some delicious memories of good times, good friends, shared moments, and people whose essence you admire and cherish.

The first word I`d pluck (see how contagious your vast knowledge of agriculture is) from my limited repertoire to describe you would be .... respect.

You practice it, you embody it, you represent it. And it's probably the best antidote for the fury and the frenzy that seems more and more inescapable every day.

The recent turnout to honor you is vivid testament to the esteem and affection people have for you. Please always remember that I'm proud to possess a lifetime membership in that club.

I don't care if the Stork's bombsight was out of line and you were unlcky enough to land where you did instead of in Sweden. God must have his reasons.

Wally

11/4/94

Wally Phillips

Dear Orion:

I apologize for the necessity of waiting till the dust-settling days are out of the way.

This note should have been at your door on January 26th, the day after you took over the Lone Ranger role at my farewell and helped me so much.

It was one of your infrequent days off and you gave it up for a friend.

Also I'm sure that with the pressure and the pace of your personal world you cherish a little time and a little quiet in whatever respite you can find.

None of which prevails in your case when you think someone needs a lift.

It reaffirmed an obvious fact. Treating those whose world you encounter with dignity and respect and concern is at the top of your priority list.

I thank you again for your kindness and your friendship. I wish I could wave a magic wandt and introduce security and joy and good times in all the things you undertake.

The nature of the person you are deservs all of the above.

I'll always be grateful.

Wally

2/23/98

WGN
Radio720
BOB COLLINS

O:

You really are one of the few people that matter a lot in my life.

The very best

Bob

June 28, 1990

Excerpt from PAUL HARVEY NEWS ... June 26, 1990

Agribusiness:

Tomorrow evening in Washington, D.C. Agriculture Secretary Clayton Yeutter will lead a salute to Orion Samuelson and his United States Farm Report. Nobody has done more to inform the farmer and to enlighten the rest of us about farming ... than has my dedicated colleague Orion Samuelson.

Paul Harvey News

May 3, 2001

Mr. Orion Samuelson
WGN RADIO
435 N. Michigan Avenue
Chicago, Illinois 60611

Good morning, Orion ...

I finished reading your profile in The Lutheran, feeling good -- reassured -- less alone.

I wish our squirrel-cage schedule allowed more time for fellowship but when one prays for guidance and new doors of professional opportunity keep opening -- I guess we're supposed to keep on keeping on.

If few understand that obligation, I know you do.

So ...

Onward and upward!

Paul

PH:jw
P.S. Again, congratulations on your Lincoln Award. Also, I hope you can find time to read in U.S. News (May 7 issue) "The Broken Heartland."

PAUL HARVEY NEWS
333 North Michigan Avenue
Chicago, Illinois 60601

GEORGE BUSH

March 2, 2009

Dear Orion,

Word has reached me that you will celebrate your 75th birthday on March 31st. Congratulations!

I will be 85 in June, and I can tell you, without reservation, that being a mere 75 doesn't hurt a bit.

Have a wonderful birthday, and best wishes for many happy returns.

Sincerely,

Mr. Orion Samuelson
c/o 5109 Farmington Close
Rockford, Illinois

P.S. I understand congratulations are also in order as you approach your 50th year in the broadcast business. Keep up the good work.

EARL L. BUTZ
Dean Emeritus of Agriculture
Room 586 Krannert Building
Purdue University
West Lafayette, Indiana 47907

Office 317-494-4307
317-494-4304
Home 317-743-1097

September 27, 1985

Mr. Orion Samuelson
Farm Caster Par Excellence
WGN
2501 Bradley Place
Chicago, Illinois 60618

Dear Orion:

A quarter century --!

No less!!

And still going strong!

If you keep on gaining momentum, at the half century dinner, you'll be the Billy Graham of Agriculture --

-- or the Willie Nelson??

Just remember Orion, "you can't use the runway behind you".

It's only on the runway ahead that we can dream, can achieve, can serve.

Orion, you have always been an inveterate optimist and a high achiever.

Last month at the local fair, this tattoo artist was tattooing images of animals on knees and elbows. Came this attractive co-ed, "Can you tattoo a kitten on my knee?"

The tattooer replied: "I certainly can."

"How much will it cost?"

"Five hundred dollars."

"Five hundred dollars?", the co-ed exclaimed, "for a little kitten, that's preposterous."

"Now look here, young lady", the artist replied. "A kitten is a difficult animal to tattoo. Tell you what - I'll make you a special price on a giraffe."

Keep young, Orion.
Keep looking up.
Set your goals high.
May health and happiness pursue you in the days ahead.

Sincerely,

Earl Butz

Earl L. Butz
Dean Emeritus of Agriculture

ELB/mb

NATIONAL RURAL ELECTRIC COOPERATIVE ASSOCIATION

1800 Massachusetts Avenue, N.W. Washington, D.C. 20036/202-857-9500

September 19, 1985

Orion Samuelson
WGN/WGN-TV
2501 Bradley Place
Chicago, IL 60618

Dear Orion:

My congratulations on 25 years of service to American agriculture.

Since 1960 your voice from WGN has been for millions of farm families a familiar welcome companion on the tractor, in the barn and around the kitchen table. Your news reports on radio and television have become a fixture in the daily routines of farmers and ranchers around the country.

As former Secretary of Agriculture, I know firsthand how important the "U.S. Farm Report" is in getting the story of agriculture to the public. It has been my personal pleasure to face your camera and microphones many times. You are a tough interviewer and always fair.

The farm broadcaster holds a unique position in the news media and you have come to exemplify the best of this breed by your professionalism and understanding and concern for the American farmer.

For millions of rural people, you have become more than a face on the TV screen or a voice on the radio, you have become a trusted friend.

As the executive vice president and general manager of the National Rural Electric Cooperative Association, I want to express my appreciation for your knowledgeable support of the rural electric program over the years.

In the difficult economic crisis facing agriculture, I know that your voice will continue to serve agriculture. We need you now more than ever in the fight to preserve the family farm concept.

For myself, the nation's 1,000 rural electric cooperatives, and the 25 million consumers we serve, thank you for 25 years of professional support of the rural electric program and agriculture.

We'll all continue to keep our dials set on WGN for your news reports for another 25 years.

Best personal regards,

Bob Bergland
Executive Vice President

BB:jb

WLQ

Sept. 13, 2007

Mr. Orion Samuelson
WGN Radio
435 North Michigan Avenue
Chicago, Illinois 60611

Dear Orion:

During the heavy rains a few weeks ago, I heard on WGN that the Kickapoo was flooding a large area.

Needless to say, I was reminded of our dedicated pursuit of the young man from the famed Kickapoo Valley!

Orion, our then WGN Farm Director, who thought more about politics than agriculture, was leaving to join the Kennedy campaign team.

I instructed our program people to find an experienced young man with a good, solid record with the farm community, advertising agencies handling agricultural accounts and a person fully "welcome" in the total rural Midwest!

I recall our first meeting at our then new broadcast center. We shook hands and what a triumph for the most precious call letters in broadcasting!

Needless to say, Orion, I hear you many times a day! Over these years, you have maintained perfectly a mighty fine voice and the essential skill of "centering" your words! As it is in professional acting, it is a "must" in announcing; and Orion, you have perfected the "art"!

Orion, my everlasting gratitude and my heartiest congratulations for a career of excellence that we want to continue for many years to come!

My warmest wishes to you and your wife, and again, thanks a million, dear friend!

Very sincerely,

Ward L. Quaal

The Ward L. Quaal Company
One Northfield Plaza
Suite 300
Northfield, Illinois 60093

Ward L. Quaal
President

Telephone
312/644-6066
Fax:
312/644-3733

November 12, 2003

Mr. Orion Samuelson
Agricultural Services Director
WGN Radio
435 North Michigan Avenue
Chicago, Illinois 60611

Dear Orion:

Again, my heartiest congratulations on being named to the Radio Hall of Fame.

I was traveling in the West but heard all but a few minutes of the one-hour show over two different stations.

Your remarks were excellent, and the entire broadcast made me very proud of radio and its greatness, today as in yesteryear.

I was pleased that Paul Harvey introduced you. All in all, it was a memorable evening that reflected great credit upon the broadcasting profession.

Kindest personal regards!

Very sincerely,

Ward L. Quaal

WLQ/rlc

ARLINGTON PARK RACECOURSE

Richard L. Duchossois
Chairman

August 17, 2003

Mr. Orion Samuelson
WGN Radio
435 N. Michigan Avenue
Chicago, IL 60611

Dear Orion,

Judi and I wish to extend our congratulations on being selected to be inducted into the Radio Hall of Fame. Your name will be amongst the wonderful names already inducted, however yours will stand out on top. All of us who have the privilege of knowing you are very gratified that you have been recognized in this magnificent way.

With kindest regards,

R.L. Duchossois

P.O. Box 7, Arlington Heights, Illinois 60006-0007 USA • Phone 847/385-7500 • Fax: 847/385-7255

PRESIDENT
GEORGE W. BUSH

September 6, 2004

Mr. Orion Samuelson
2130
Northbrook, Illinois

Dear Orion:

It was good to see you in Alleman last week. The event was a big success, and I appreciated you taking the time to be there.

We are marching to victory, and I am proud to have you on my team.

Laura joins me in sending our best wishes.

Sincerely,

George W. Bush

NOT PRINTED AT TAXPAYERS' EXPENSE
PAID FOR BY BUSH-CHENEY '04, INC.

RONALD REAGAN

October 8, 1994

Dear Friends,

Greetings to all those gathered tonight to honor Orion Samuelson -- a great radio man of the ages, an important voice for agriculture, a legendary newsman and a dear friend to all of us.

I wish there was some way my schedule would have allowed me to join you tonight, but it wasn't possible. One of these days I'm hoping "retirement" will kick in! I wanted to be there so I could return a ten year old favor. You see, back in 1984, it was Orion Samuelson who emceed a birthday party for me in Dixon, Illinois. Orion made sure Mrs. O'Leary's cow wasn't around to kick over my cake, sparing Dixon from the same fiery fate that Chicago suffered. Of course, the cake could have easily toppled over without any help from a cow due to the number of candles it was balancing on top. I'm not going to reveal how many candles there were -- but I think it's safe to say that a fire extinguisher was close at hand just in case. Nevertheless, since that party I always felt like I've owed him one and I'm delighted to send my warmest greetings on this special occasion in his life.

As all of you know, Orion has been an important voice for agriculture for a great many years -- four decades, actually. He's been a source of news and information -- from our own backyards to faraway lands in the heart of what used to be the Soviet Union.

Yes, Orion was there, back in 1983, when he accompanied my Agriculture Secretary, John Block, to Moscow to sign a long-term grain agreement with the Soviets. Orion saw to it that the news of such a successful effort made it home. He informed us of this breakthrough affecting our crop sale to a foreign market -- a victory which expanded trade opportunities for American agriculture. This was great news for our farmers and Orion was the right journalist to deliver it. There have been many other important moments and he was right there to bring the news to the people.

-2-

So on this occasion, Orion, I'd like to commend you for all you've done for agriculture. As a former broadcaster myself, I know it's not a business for the faint hearted. Looking at the success you've had at W.G.N., and the remarkable career you've had, I'm glad I got out of radio when I did and went into less competitive fields, like acting and politics!!

Nancy joins me in sending our best wishes for a wonderful evening. God bless you.

Sincerely, Ronald Reagan

DEPARTMENT OF AGRICULTURE
OFFICE OF THE SECRETARY
WASHINGTON, D.C. 20250

September 13, 1985

Mr. Orion Samuleson
Vice President
WGN Continental Broadcasting Company
2501 Bradley Place
Chicago, Illinois 60618

Dear Orion:

My compliments and congratulations to you on completion of 25 years of service with WGN.

WGN has been a broadcasting giant because of talented individuals like you giving it punch as your shows echo from the Atlantic Ocean to the barbed wire fences of south Texas.

Most memorable to me, of course, was your trip with me to the Soviet Union in 1983 when we signed that long-term trade agreement with the Soviets.

My best to you always.

Sincerely,

John R Block

JOHN R. BLOCK
Secretary

U.S. House of Representatives
Committee on Agriculture
Room 1301, Longworth House Office Building
Washington, DC 20515

June 21, 1990

Orion Samuelson
WGN/Tribune Radio Network
435 N. Michigan Avenue
Chicago, IL 60611

Dear Orion:

It is with great pleasure that I congratulate you on the upcoming 15th anniversary of your television show "U.S. Farm Report."

For the past 15 years, "U.S. Farm Report" has played a very important and valuable role in agricultural communications. "U.S. Farm Report" puts into perspective for a nationwide farm audience the events and trends affecting agriculture here at home and abroad. Equally important is the fact that your show provides the American public -- both farmers and non-farmers -- a better understanding of U.S. agriculture, its complexity and uniqueness, and its impact on people's lives.

Since becoming Chairman of the House Agriculture Committee, I have been interviewed for "U.S. Farm Report" countless times. I have always been impressed by the professionalism and knowledge you and your staff have. The American farmer and the issues surrounding agricultural policy are better understood by the American public because of your work on "U.S. Farm Report."

American agriculture is indeed very fortunate to have a person of your caliber and dedication serving its cause. I wish you and the staff that produces "U.S. Farm Report" continued success in your activities.

With warmest regards,

Kika de la Garza

E (Kika) de la Garza
Chairman

DLG/jrd

JESSE HELMS
NORTH CAROLINA

United States Senate
WASHINGTON, DC 20510-3301

June 27, 1990

The Honorable Orion Samuelson
The Voice of America's Farmers
U.S. Farm Report
Chicago, Illinois

Dear Orion:

A lot of guys get "The Honorable" treatment as a matter of course, but you got it the old-fashioned way: You earned it.

You're not just a broadcaster -- you're an institution. And it is good that you and U.S. Farm Report are being deservedly recognized on the 15th anniversary of the beginning of this fine service to everybody interested in agriculture.

As an old (very) former broadcaster, I congratulate you and extend my best wishes for many more decades of splendid service.

Sincerely,

Jesse Helms

JESSE HELMS:pd

THE SECRETARY OF AGRICULTURE
WASHINGTON, D. C. 20250

June 27, 1990

Mr. Orion Samuelson
"U.S. Farm Report"
WGN Continental Broadcasting Corporation
Tribune Building
Chicago, Illinois 60611

Dear Orion:

Congratulations on the 15th anniversary of the "U.S. Farm Report." I am proud to have been part of your programming for the past two decades.

Anyone who has followed the "U.S. Farm Report" over the years would agree with me that it is one of the most respected farm broadcasts in the country. As a farmer, I know the importance of having easy access to reliable, up-to-date news on markets, the farm community and the Washington-perspective on agricultural issues. The "U.S. Farm Report" has risen to that need for the last 15 years, and continues to do so with high journalistic standards and admirable professionalism.

Orion, you are tops in your profession and have made a major contribution to America's farmers and ranchers over the years. I speak for all your fans at the U.S. Department of Agriculture--best wishes to you and the crew of the "U.S. Farm Report" for another 15 years of superlative broadcasting.

Sincerely,

Clayton Yeutter

CITY OF CHICAGO
CHICAGO CITY COUNCIL
ALDERMAN OF THE 42ND WARD
ROOM 300 CITY HALL

ALDERMAN BRENDAN REILLY

TELEPHONE: (312) 744-3062

September 22, 2010

Orion Samuelson
WGN Radio
435 North Michigan Avenue
Chicago, IL 60611

Dear Mr. Samuelson:

I am writing to congratulate you for 50 years of service as the Agribusiness Director to WGN radio and for hosting WGN Radio's agriculture program. While I would have very much liked to attend tonight's celebration in Rosemont, I had a previous commitment that unfortunately prevented me from joining you.

Please accept my most sincere regrets – I wish I could have joined you for this special occasion. As a token of my appreciation of your service to the agricultural community in the Midwest, I have introduced legislation for honorary designation of a portion of East Illinois Street as "Orion Samuelson Way."

The street sign will be installed in the near future and I look forward to the dedication ceremony to celebrate your contributions to both radio and agriculture.

Sincerely,

Brendan Reilly,
Alderman, 42nd Ward

NOT PAID FOR AT TAXPAYER EXPENSE

A resolution

adopted by The City Council
of the City of Chicago, Illinois

Presented by ALDERMAN BRENDAN REILLY on DECEMBER 8, 2010

Whereas, Orion Samuelson is heard on WGN Radio, where he has served as Agribusiness Director since 1960. Orion presents daily agricultural and business reports on WGN, in addition to co-hosting The Morning Show, *WGN Radio's agriculture program, with his longtime colleague Max Armstrong on Saturdays at 5am*. Orion is also heard daily on stations across the Midwest with his syndicated National Farm Report and Samuelson Sez. Orion and Max are seen weekly on rural channel RFD-TV, carried on Dish-TV and DirecTV on This Week in Agribusiness; and

WHEREAS, His life-long commitment to agriculture has been recognized by organizations in all segments of agri-business. In 1985, Orion was inducted into the Scandinavian-American Hall of Fame. He has also been inducted into the National 4-H Hall of Fame and the NAFB Hall of Fame; and

WHEREAS, Orion was named a Laureate of the Lincoln Academy of Illinois and received the Lincoln Medal, the highest award bestowed by the state of Illinois, shortly thereafter, the University of Illinois presented Orion with the Honorary Degree of Doctor of Letters. In October of 1994, he was honored as "Man of the Year" by Heifer Project International on its 50th anniversary; and

WHEREAS, At the 1997 Illinois State Fair, Governor Jim Edgar changed the name of the Junior Livestock Building to the Orion Samuelson Junior Livestock Building as a tribute to Orion's nearly four decades of service to the agricultural youth of Illinois; and

WHEREAS, In November 2003, Orion received the highest award in the radio broadcasting industry when he was inducted into the National Radio Hall of Fame in Chicago. Hall of Famer Paul Harvey presented Orion with the award on the national broadcast hosted by Larry King; and

WHEREAS, The Honorable Brendan Reilly, Alderman of the 42nd Ward has informed this august body of this significant milestone; now, therefore,

BE IT RESOLVED, That we, the Mayor and the members of the Chicago City Council assembled here this day of the eighth of December, 2010 does hereby join with the family and many friends of Orion Samuelson in celebrating his accomplishments, the renaming of a portion of East Illinois Street, and his exemplary life; and

BE IT FURTHER RESOLVED, That a suitable copy be prepared and presented to Orion Samuelson and his family as testimony to his wonderful life.

Richard M. Daley
MAYOR

Miguel del Valle
CITY CLERK

A Sunday morning blizzard in February left the roads of western Wisconsin drifted shut. Lena decided to stay home with Little Ole, but Ole managed to find a way to get to church in time for the 11 o'clock service. The only other person there was Pastor Ingvoldstadt.

After waiting a few minutes to see whether anyone else would show up, the pastor said, "Vell, Ole, I guess we should yust go home. Dere's no sense in my doing a service yust for vun person."

Ole said, "Ya know, pastor, even if only vun cow showed up at feeding time, I would feed it."

On the set of "This Week in AgriBusiness."

Chapter Twenty-Two

Samuelson Sez

Over the years, I've worked very hard to keep news stories in my market reports and newsmaker interviews objective. If opinions are expressed, I let my interview guests do it. For example, when I did an interview with Charlie Stenholm, the retired longtime Texas congressman who served on the House Agriculture Committee, he talked about the unwanted horse problem. Until PETA and the Humane Society of the United States got involved, we had three processing plants where horses were slaughtered and their meat was shipped overseas where people eat horse meat much the way we eat beef and pork. PETA and HSUS, using emotional and flawed arguments, got laws passed that drove those plants out of business. So we wound up with about 70,000 unwanted horses a year and when people can't afford to feed them, they turn them loose. Those horses eventually starve to death, a much worse death than the humane treatment they would have received in a packinghouse. Well, Charlie Stenholm came out strongly saying, "We've got to change this. We've got to change this, because you're condemning horses to a terrible death, and the law is wrong." So, I let him do the editorializing. I asked the questions and yes, I have a strong prejudice on this, so I'm sure I asked the questions in a way that would get the response that would reflect what I feel, but in Charlie's case, he didn't need any guidance from me.

There are times when I feel the need to give my opinion and for those times, I use *Samuelson Sez*. I started it in the 1980s and once a week, I tell how I feel about an issue of importance to agriculture. *Samuelson Sez* serves the same purpose as the editorial page in a newspaper or in farm magazines. It gives me an outlet for my opinion and allows me to do a better job of keeping editorial bias out of news stories. For example, when I report a story on an animal identification program and I say that so many ranches and farms have signed up and others are refusing to do so for this reason or that reason, that's a news story. But, when I say, "I think we ought to do it," then that's an editorial, and I'll put that in *Sam Sez*.

It starts with the TV show. I ad-lib my opinion and we'll take the audio and put it on our Saturday morning show on WGN Radio as well as on about 120 other radio stations that carry it in syndication. Around 2004, a syndicator of agriculture newspaper columns approached me. So I'm in about 30 rural newspapers, weekly papers, mostly.

I've been surprised a few times by the response to my opinions. The animal ID program I mentioned above is one example. I'm for it. Every livestock operator should have to register his herds and when there's an outbreak of disease, the national ID program would allow us to react quickly, find the source and reassure the public that the meat they buy in the store is safe. Instead of every producer getting shut down and everybody losing money, if you had animal ID, you could go right to the source and put a stop to it. Well, the reaction from some producers was anger, both at the idea and at me. I had producers telling me, "You want Big Brother to watch us and control us. You want to know where we are so that we can be taxed more," and that sort of thing. It was not nice. But, I also heard from a lot of people who said, "You're absolutely right. It will benefit all of us." And, by the way, there is a precedent. If you grow carrots for Gerber Baby Foods, you have a contract that says every acre of carrots has to be carefully tracked: what day you planted it, what you put on that ground and when. Everything, so that if there is a contamination issue in Gerber's baby carrots, they can go right back to the very field and immediately get the thing corrected. So it's not something that's new to agriculture, but boy, the livestock people didn't like it — and don't like it.

Most of the time, I think my opinions are in sync with producers, but now and then I'll get e-mails that question my relationship to my mother and all that sort of thing, some pretty strong comments on how wrong I am.

One of my favorite *Sam Sez* letters arrived in 1979 from former Secretary of Agriculture Earl Butz. At that time, he was Dean Emeritus at Purdue University. In a style uniquely his own, Earl "roasted" me with a letter of congratulations on 25 years in farm broadcasting.

As I say on *Samuelson Sez*, my mission is to stimulate thinking. You don't have to agree with me. I don't expect that, but if I can get you thinking about an issue, then maybe that leads to discussion, and that could take us

EARL L. BUTZ
Dean Emeritus of Agriculture
Room 586 Krannert Building
Purdue University
West Lafayette, Indiana 47907

Office 317-493-2223
317-494-3212
Home 317-743-1097

Mr. Oracle Samuelson
Bumbling Broadcaster
World's Greatest Nuisance
Cackling Chicago,
Ill Noise

August 10, 1979

Dear Oracle:

So you've been fanning the breeze for over 25 years!

Hot breeze, that is!

And still - the farm problem goes on and on!

Perhaps it's stirred by hot breezes of the Bumbling Broadcaster over the World's Greatest Nuisance.

If the Midwest can survive another 25 years of ORACLE OUTRAGE (alias Samuelson Sez), the Tractor Treks of 1979 will seem tame beside the Helicopter Hops of Fuming Farmers in 2004.

And the World's Greatest Nuisance will have its camera there to record the age-mellowed voice, the silver-thatched crown, the weather-wrinkled face, and the still convincing conversation of Orien the Oracle as he chronicles the confrontation.

And, in a more serious vein, Orien will be doing his always great job as America's Number One Farmcaster.

However, 25 years hence, Orien may find it a little difficult to remain fully active on all three fronts:

Wine, Women and Song -
He will give up singing!

I join in the salute to a:

Superb Farmcaster -
Loyal American &
Personal Friend.

With every good wish for another 25 years, I am

Sincerely,

Earl

Earl L. Butz
Dean Emeritus of Agriculture

to an agreeable solution.

What follows is a sampling of more recent *Samuelson Sez* covering a wide variety of topics. They are grouped by issue. While some of the numbers may be out of date (remember $3 corn?), it's interesting to see how some topics remain at the forefront over a period of years and, in a few cases, how my opinions have changed over the years. Feel free to let me know how wrong I am! I've also scattered several more photographs of people, events and travels.

- Defending Agriculture
- Defending the Meat Industry
- Dairy Farming
- Food Prices
- Prices Don't Go the Same Direction Forever
- Ethanol
- Getting Along with Neighbors
- Telling Agriculture's Story
- Mad Cows
- Don't Call it Swine Flu!
- Animal Tracking
- If You Eat, You're Involved in Agriculture
- PETA, HSUS and Other Nonsense
- Unwanted Horses
- The Environment and the Agencies That Are Supposed to Protect It
- Food Police
- Checkoff Programs
- Ag Exports
- Corporate Farming
- County and State Fairs
- Farm Safety
- Estate Planning
- Keeping Rural Life Livable
- The Church
- Politically Incorrect
- Politics and Government
- Banks and Bankers
- USDA

DEFENDING AGRICULTURE

Support Ag In The Classroom - October 8, 2007

I have been a supporter of the Ag In The Classroom program for many years, and for several reasons. The two most important are the need to get agricultural education, particularly education about animal agriculture, into the classrooms of fourth-grade students in towns and cities across the country. Secondly, the program works! It does tell the story of agriculture in a very positive way, to counter-balance the claims of organizations such as "PETA," People for Ethical Treatment of Animals, that are generally based on emotion, not sound science. So I encourage you to support Ag In The Classroom; to start that education at an early age, and help people understand the role of animal agriculture in the food chain.

The reason I bring it up now is because of a happening in an elementary school about 10 miles from where I live. A 44-year-old art teacher became a vegetarian about a year ago and decided it was time to tell his students that vegetarianism was "the only way to go." He gave material to his students in the classroom that depicted any foods from animal agriculture as poison and dangerous to their health and encouraged them to take the material home and share it with their family.

He said he felt he had the right to do this because there were posters in the school cafeteria promoting the health benefits of milk and dairy products and the "Three-a-Day Dairy" program. When school officials refused to remove the posters, he said he felt compelled and within his rights to warn his students of his perceived health danger in consuming dairy products. He was angered by the school decision, decided to take things into his own hands, and instead of teaching art full-time, spent part of his classroom time teaching his students to become vegetarians.

It didn't take long for parents, who felt it was their responsibility to teach their children good eating habits, to register their complaints. They told school officials that teaching vegetarianism on school time was wrong and the teacher should be instructed to stop and stick to teaching art. When he refused, he was first suspended, and then on a 7-0 vote by the school board was dismissed.

I commend the School Board for its action, but incidents like this simply reinforce the need for the Ag In The Classroom program. I am delighted by the involvement of farm organizations who financially support the program. And I am especially pleased with farm families who establish personal contacts with classes in city schools with videotape of their farming operations and classroom visits to explain what they do and what it takes to put wholesome nutritious food on the dinner table of those students.

I urge you to support Ag In The Classroom.

My thoughts on *Samuelson Sez.*

Beware of False Rumors on Internet - November 11, 2007

There is no question that the Internet has played a positive role in the lives of all of us, bringing information from around the world to our fingertips, instantly. But along with the positives, there are some negatives and this is one that really bothers me. Anyone can put anything on the Internet, truthful or not, and people will believe it.

One of my favorite weekly country newspaper editors, Karen Parker, publisher of *The County Line* in my hometown of Ontario, Wisconsin, described the Internet this way in a recent column: "The Internet is a wonderful mechanism that has turned commerce on its head, offers information at the click of a mouse. It allows people to communicate easily and cheaply, but it

For decades, whenever asked and whenever possible, I have spoken enthusiastically about the great work that American agriculture is doing to feed the world and about the problems facing our producers.

is overrated as a force for world peace or even local peace. People without the slightest credentials or even minute facts can randomly send reams of falsehoods around the world in a matter of seconds. Hence, we have total absurdities, such as individuals contracting AIDS from contaminated needles buried in movie theater seats; government proposing a tax on e-mails; or plants dying after being fed microwaved water; all presented as the 'gospel truth.' We get crooks in Nigeria defrauding people out of their life savings; access to miracle drugs whose only miracle is making your money disappear and cranks and kooks posing as journalists." That's Karen Parker, publisher of *The County Line* weekly newspaper.

The reason I bring it up this week is because there is just such an e-mail now circulating on the Internet, attacking the quality and safety of food at McDonald's.

The writer of the e-mail who says he (or she) is a cattle producer states that McDonald's buys diseased and "downer" animals at stockyards and they become part of the beef that goes into their hamburgers. Because the writer knows that "to be absolutely true" he urges people to stay away from McDonald's, adding that several members of his family became ill after eating at McDonald's several years ago.

I do not work for McDonald's, but that e-mail is pure nonsense. All you need do is look at the facts. McDonald's is the No. 1 hamburger-seller in the world. Just ask this question. Why would the company endanger its reputation and its bottom line by doing something that stupid?

I don't know the e-mail writer, but I do know the lady who is the Vice President of Quality Worldwide for McDonald's and she told me there is no truth to the claims in the e-mail. She said, "No way would McDonald's endanger its reputation nor, more importantly, endanger the health of its customers. We only buy quality USDA inspected beef." She grew up on a farm in northern Illinois and was a 4-H Club member. I will believe her. I will not believe an anonymous writer on the Internet. Just because it's on the Internet is no reason to accept it as gospel truth. Check it out!

My thoughts on *Samuelson Sez.*

Challenge Facing General Farm Organizations - January 19, 2008

During my recent visit to New Orleans this month to cover the 89th annual meeting of the American Farm Bureau, I was reminded again of the growing challenge facing a general farm organization. That challenge is trying to satisfy policy needs of a diverse commodity producer membership that produces milk, citrus, pork, beef, grain, vegetables, fruit and dozens of other agricultural commodities. How do you adopt policy that will satisfy all of those producers?

Perhaps the biggest division at the moment is between grain producers and farmers and ranchers in the livestock business. Grain producers are enjoying record high prices because of strong demand for energy as well as strong demand in the export market. But those high feed prices translate into tough times for livestock producers and they are not happy campers right now. I will attend the National Cattlemen's Beef Association annual meeting in Reno in February and I know I'm going to hear cattlemen say the tax break for ethanol should be eliminated to help level the playing field for them.

There was a time when farm policy was the domain of the three major farm organizations... National Grange, National Farmers Union and Farm Bureau; later they were joined by the National Farmers Organization and the American Agriculture Movement. All five are general farm organizations.

But then, in the last four decades we have seen a proliferation of commodity-specific policy organizations representing dairy, hog, beef, corn, wheat and other producers. Leaders of these commodity organizations can focus on specific issues that affect only their producers. The National Corn Growers Association can lobby for a continuation of the tax break for ethanol; the NCBA can lobby to kill it; Farm Bureau would find it impossible to adopt either of these policies.

The discussion I heard among Farm Bureau leaders basically said, "Give it some time, farm prices always moderate and all of us will find a way to work through it." AFBF President Bob Stallman summed it up this way: "Once again, farmers and ranchers from all across the country, who raise a range of crops and livestock, have come together to decide what is best for U.S. agriculture overall." It will be interesting to hear what cattlemen in Reno and grain producers at the Commodity Classic in Nashville have to say about the feed vs. fuel issue.

I think we still need general farm organizations and I wish success to their leadership as they deal with policy for a diverse U.S. agriculture.

My thoughts on *Samuelson Sez.*

I Am a Cheerleader - February 15, 2008

Am I a cheerleader for American agriculture? You bet I am, and I am proud of it. Some responses to my recent *Samuelson Sez* on "sound science vs. emotion" causes me to put that cap on today to cheer lead for America's

John Deere began expanding into the United Kingdom in the mid-'60s. This interview was during one of our trips in the 1970s.

farmers and ranchers

There were some lines in two or three e-mails that I found extremely irritating. Let me share a couple with you. One who took strong exception to what I said about sound science and my comments on the use of rBST in the dairy industry said, "Well, dairy farmers are greedy. They want to produce more milk to make more money and don't care about the health of their cows or us consumers." This same person then went on to say, "Crop farmers and all livestock farmers are greedy. They use this technology without any consideration of consumers, so they can make more money."

I do agree on one point; in a capitalist system most of us do have a goal of putting more money in the bank and for many, including farmers and ranchers, that is a daily struggle. But greed and a lack of caring for the well-being of people and animals are two criticisms I will not accept. Let me share some of my thoughts on the people who put food on my table, clothes on my back, a roof over my head and now, fuel in my gas tank.

First of all, less than 2% of our population does just that for the rest of us in this country, so the other 98% of us are free to pursue other careers and activities because we don't have to worry about our food supply. Secondly, when farmers make money, they spend it. All you need to do is look at that recent quarterly earnings report from Deere and Company, a 55% increase in profit because they sold more farm equipment. When farmers and ranchers make money, they put it right back into the economy, providing jobs for the people who make those machines and the many others who provide services to agriculture.

Finally, farmers are the nation's top conservationists and environmentalists. In a profession where there is no such thing as an eight-hour work day, they work hard to preserve and improve the land, to make it productive for the generations to come, and we need to recognize them for that effort.

Greedy? NO! Conscientious, hard working, taking pride in what they do? YES! So I will not accept that kind of criticism from people who disagree. Disagree with me on sound science if you like, but don't level unfair criticism on the American farmer who feeds you.

Some early thoughts ahead of National Agriculture Day in March on *Samuelson Sez.*

Critics Want A Meatless World? - May 17, 2008

The subject of this week's *Samuelson Sez* was prompted by an e-mail from a farmer who asked me this question: "Why are people in the United States so angry with those of us in farming and ranching? They are blaming us for everything; for taking feed and food and putting it in our gas tanks.

December, 2008 at Spike O'Dell's retirement party at Chicago's iconic Billy Goat Tavern.

They are blaming us for using technology. They are blaming us for mistreating animals. We seem to get blamed for all the ills of the world, and frankly, I wonder if these people understand that if my neighbors and I weren't out here using the technology and doing what we do, there would be no food on the table, or if it was there, it would cost a lot more. There would be no clothes on their back, no roof over their head and yes, no energy in the tank."

Well, I hear you my friend and I share your frustration. But the criticism is there and it seems to be building. I was looking at the website of *The Indianapolis Star* newspaper last week and in the Letters to the Editor column, I found this letter written by Ike Gallego of Indianapolis, a reader who obviously would like to put all livestock producers out of business.

This is what he wrote... "The prestigious Pew Charitable Trusts and Johns Hopkins Bloomberg School of Public Health recently concluded that factory farming takes a big toll on human health and the environment, undermines rural economic stability and fails to provide humane treatment of livestock. Capping a two year study, the report calls for a national phase-out of all intensive confinement of farm animals."

The letter writer goes on to say, "The report is long overdue. For the past 60 years animal agriculture has been devastating our country's vital natural resources, including soil, water and wildlife habitats. It has been generating more greenhouse gases than transportation. It has been elevating the risk of chronic diseases that account for 130 million deaths annually. It has been abusing billions of innocent animals." And here is the final line: "The only long-term solution to this tragedy is to gradually reduce the consumption of animal products to zero." In other words, no more meat on anyone's dinner plate; we should all become vegetarians for the good of the planet.

Of course, it is distorted, untrue and unfair, but it is out there and we in agriculture need to deal with it. Perhaps one of the reasons it's out there is because farmers and ranchers do their job so well. We don't have food shortages in this country and, despite the recent food price increase, U.S. consumers still spend less money for food than consumers in any other country. It is more important than ever that agriculture responds with a unified voice to help people understand that without agriculture, there would be nothing.

My thoughts on *Samuelson Sez.*

Animal Agriculture's Battle to Survive Needs Help - April 3, 2009

Last month at the 100th anniversary convention of the American

Feed Industry Association, I was asked to address the membership. My assigned topic, "Challenges Facing the Feed Industry in the Next Century."

It was an easy assignment because at the top of my challenge list for the animal feed industry is the challenge from animal rights organizations to eliminate animal agriculture in the United States. If they are successful in doing that, they will in effect, eliminate the feed industry as well; an industry that has served livestock and poultry producers for more than a century.

It was a timely topic because at the meeting, the convention issue of *Feedstuffs*, the weekly newspaper for agribusiness, had a front page story entitled "Proposition 2 Opens the Door to Push for Veganism;" the article written by Rod Smith stated, "California's passage last November of the Ballot Initiative on Farm Animal Housing, Proposition 2, has opened the door for animal rights activists to begin promoting a vegan lifestyle for California consumers; that according to a coalition of activist groups. The coalition said it was establishing the Operation Prop 2 Follow-Through Campaign, with the intention to promote a vegan diet in California. That coalition is led by the Farm Animal Rights Movement, FARM. It said it wanted to take advantage of the extent to which Proposition 2 made people aware of 'Factory Farm Atrocities' and take the awareness to what would be the next level, to provide animals complete protection from those atrocities by urging Californians to become vegetarians; which means not only not consuming food and beverages produced from animals but also not wearing anything produced from animals, such as leather or wool."

If you think this challenge isn't real, think again. When you have organizations like the Humane Society of the United States with $200 million in assets, and one of its grass-roots coordinators recently saying, "My goal is the abolition of all animal agriculture," this is a real threat. Add to that, PETA, People for the Ethical Treatment of Animals and its ability to make headlines and to flood elementary schools with misleading literature about the treatment of animals by farmers and ranchers and you are dealing with well-planned, well-financed campaigns. Those of us who believe in the animal agriculture industry of the United States must come together to find the dollars and speak with one voice to tell consumers the truth. It is time to get involved!

My thoughts on *Samuelson Sez.*

Continued Attacks on Animal Agriculture - February 10, 2010

Livestock producers, like it or not, you continue to be in the national media spotlight and most of that coverage is not very positive and not very

2010 in the WGN Showcase Studio on Michigan Avenue with Dave Eanet and Greg Jarrett. That's the back of engineer Bob Ferguson's head, and in the background behind the glass is morning show producer Jim Wiser.

flattering. If you thought 2009 was a tough year for animal agriculture attacks with the *TIME* magazine article and other organizations leading the way, 2010 is off to an even tougher start.

So far, in the first two months of the year, Katie Couric, CBS television news spent considerable time talking to what she termed is a major problem... the overuse of antibiotics in livestock that eventually turn up on our dinner tables. ABC's Diane Sawyer provided television coverage featuring video that was shot on a dairy farm in New York showing tail docking and very inhumane treatment of dairy animals and of course California has already banned tail docking in that state.

But the attacks continue and is it part of an organized campaign? Yes, I think so. We are going to have to find quicker, better ways to respond, and we must also keep in mind that if you know a neighbor that is mistreating livestock or poultry in any way, then do something about it because those cell phone cameras can be carried anywhere and concealed, but that video shows up on national television.

Now some livestock producers did get some good news a few days ago. Secretary of Agriculture Tom Vilsack said, "We're going to stop pursuing a national animal ID program." He said that it's going to be turned over to the states; the Department of Agriculture will pay most of the funding and it will not affect those animals that are in back yard flocks or livestock that is sold and consumed locally. It basically will involve animals that are involved in interstate commerce. Again, judging from the "hate mail" I got when I supported the national animal ID program, I am sure this pleases a lot of you.

I still don't understand though, how you can favor "country of origin" labeling on one side and be totally against "traceability in animals" on the other side.

My thoughts on *Samuelson Sez.*

Farming Regulations Get Tougher - February 16, 2010

I'm grateful to you, my listeners, for keeping me posted on activities and events taking place in your part of the country. Very often, what you tell me leads to a *Samuelson Sez*, and that's the case this week.

A farmer in Pennsylvania e-mailed me and said "there are some restrictions that will be put in place in Pennsylvania on March 1st. The restrictions and requirements will apply whenever farm trucks, farm tractors and other agricultural vehicles are used on any state or township road, including just crossing the road." These are the rules:

Don Peasley was the man behind the camera for many of the older photos in this book, especially of events near his home in McHenry County, Illinois. Don began his career as a photojournalist in 1947 and recently donated his huge collection of photos and negatives, the majority related to agriculture, to the McHenry County Historical Society for preservation and display.

#1 - Anyone under 18 years of age will not be able to legally operate a farm tractor or truck, if the tractor or truck is pulling another vehicle, farm equipment or trailer, and the vehicles together weigh more than 17,000 pounds. This includes the farmer's children.

#2 - Medical certification required - Any driver on a public road of a farm tractor or truck will need to be medically tested and have a certificate declaring the driver to be physically qualified.

#3 - Driver Logs/Employer Recordkeeping required - Drivers of trucks or tractors towing implements or farm trailers will be subject to the same hours-of-service requirements of trucking companies. It will include minimum periods of "break time" limits on hours driven between breaks, along with record keeping (commonly known as drivers' logs). Farmers will be responsible for getting and keeping records of all the paperwork that federal regulations require trucking companies to keep and maintain.

#4 - Vehicle Inspection and Maintenance required - This means farmers will be required to conduct pre-trip inspections and complete written post-trip reports on the functions of the vehicle's safety equipment. Inspections and reports must be filled out each day the vehicle is used.

Again, all of the above rules apply if the combined weight of the tractor and towed equipment exceeds 17,000 pounds and if you are just crossing a state or township road! This is just what farmers and ranchers need…more paper work to produce food. The rules take effect in Pennsylvania March 1st and I would like to know if these rules are in effect or in the works for any other state. Let me hear from you at orion@agbizweek.com.

My thoughts on *Samuelson Sez.*

An Empty Plate : A Great Tactic - March 4, 2010

This week on Samuelson Sez, I would like to do "Show and Tell." Since I don't have a camera to share the picture with you, I'll try to paint a "word picture" about the contents of a package that arrived on my desk a few days ago. In the package, there was a news release and a clear glass plate, with a note attached to the plate.

Let me read the words:

"An Empty Plate. That's what the Humane Society agenda promises. HSUS deceptively uses this name to fund a PETA-inspired anti-meat campaign. Don't be fooled. HSUS isn't your local animal shelter. HSUS cares about fundraising. Wisconsin farmers care about animals. Get the facts at www.Humanewatch.org."

The note that accompanied the press release and the plate said, "This

At my WGN Radio desk in 2010.

plate and letter were distributed by corn, soybean, dairy and hog farmers to Wisconsin legislator offices in the State Capitol in Madison on Wednesday, February 23."

Well, I say "hats off" to the corn, soybean, dairy and hog farmers of Wisconsin for finding a unique way to tell the story in the press release, pointing out that the Humane Society of the United States is not your local animal shelter, it is a lobbying organization with a huge budget. Less than 4% of that budget goes to local animal shelters, while the bulk of that money goes to campaigns to shut down animal agriculture. For example, a lot of Humane Society funds went to Californians for Humane Farms, the lobbying group that effectively shut down California's poultry industry.

The press release concludes "We don't want people fooled by those who say they care about animals while attacking those of us who show we care every day. We feed animals. We feed people. We are the Wisconsin farmers."

Your response to my "How should we tell the agriculture story to non-farm consumers" a few weeks ago called for more activity by individuals, the need to be more pro-active rather than always being on the defensive and finding attention-getting ways to tell the story. The "Empty Plate" campaign does all that and more. It fits all commodity groups and in this instance, four groups worked together to make a very strong point.

It's a great idea to tell the ag story in a positive way and I suggest you consider doing the same with legislators in your state. I guess I would add just one more line to the message and that would be... No Farms, No Food!

My thoughts on *Samuelson Sez.*

It's Up to Us to Support Our Ag Programs - April 21, 2010

These indeed are grim and challenging times for officials at the state, county, township, school board level; all dealing with budget short-falls, trying to maintain important programs but finding it impossible to do so because they simply don't have the dollars to do it.

Every week I receive newspaper clippings, e-mails and letters from people in rural communities who are concerned about losing 4-H programs and Extension personnel, as well as faculty at land-grant universities. My state of Illinois is a prime example, a $13 billion budget shortfall forcing termination of four dozen County Extension Director positions and layoffs and dismissals of faculty at the University of Illinois College of Agriculture, Consumer and Environmental Sciences.

I hear from parents with students in rural high schools who are being told that the school board is considering cutting the vocational agriculture and FFA programs because they don't have the dollars to fund them. How can this be happening in a rural high school?

Now, we're not alone because there are many worthy programs and needs that are fighting for the same dollars; and as individuals, many of us are fighting a personal budget shortfall. But in agriculture if we feel strongly about the need for people who provide Extension service, whether it be to senior citizens, 4-H'ers, Master Gardeners or farmers; or if we are truly concerned about FFA programs and the education they provide young people that could lead them to any one of more than 300 careers, then I guess we pull up the bootstraps and find a way to financially support these programs without turning to government or tax dollars that aren't there anyway.

Let me give you one example of how we as individuals can do that. I am a member of the Board of Trustees of the National 4-H Council. We just concluded our semi-annual meeting in Washington, DC and there we talked about a special program; the "4-H Give 2 Vote" challenge. It is a fundraiser to benefit state 4-H programs using mobile and on-line technology. The Council hopes to rally 4-H's more than 88,000 Facebook fans in a giving competition to support their state 4-H program. Each $10 donation equals one vote. The state garnering the most votes will receive a bonus of $5,000 from J.C.Penney for 4-H work. You can get more information at www.4-H.org. This is just one of many ideas that we in agriculture can generate to support programs that are important to the rural community. Let's put on our thinking caps and get to work!

My thoughts on *Samuelson Sez.*

The Schohr Family Says it All - April 30, 2010

A few weeks ago I urged all of you to become more involved in telling the positive story of American agriculture. What a job some of you have done, particularly the Schohr family in California; brothers Ryan and Steven and their sister Tracy worked together to write an Op-Ed piece that was printed in the *San Francisco Chronicle* on April 22nd. Let me share some of what the Schohr family had to say.

"As you celebrate Earth Day today, be thankful you live in a country where you have access to affordable and healthy food. It is because of our family farming business and others like it that each day Americans have a variety of the world's best fruits, nuts, vegetables, grains and meats. We are the fifth-generation to raise cattle and rice, and our families, on the same land. We have lessons passed down from our father and grandfathers on how to work in balance with nature to produce food from the land. At the same time, we use cutting-edge technologies like GPS that help us use resources more economically and efficiently.

"Making a living from the land is not easy, but it's how we support our families and communities. As agricultural producers, we comply with ever-changing environmental regulations to ensure water that's used to produce our food supply remains clean, and that the fertilizers, feed, seed and energy necessary to grow the food are safe for the environment and your families, as well as ours. Our farming operation exists in harmony with nature. Day and night, we see the wildlife that depends on our farming operations. Species such as deer, rabbits, pheasants, turkeys and even wildflowers can be found on the rangelands grazed by our cattle. Did you know that 75% of all threatened and endangered species depend upon private lands? Our rice fields, which lie in the Pacific Flyway, provide forage to thousands of migratory waterfowl each year.

"Our parents instilled in us a love of the land, family and hard work, but also an understanding of the economics necessary to remain in business. It's important for all Americans to help keep farmers in business, which in turn helps protect open spaces, keep urban sprawl in check, promote carbon sequestration and protect wildlife habitat. It is time for those who care about the environment to join forces with the family farmers and ranchers to protect farms as well as nature."

Not many barns have a clocktower. This is the Balboughty Home Farm near Scone, Perthshire, Scotland. The gentleman to my left is the Earl of Mansfield.

What an outstanding job by the Schohr family! This is proof again that individuals can make a difference. I would suggest you take the ideas they expressed, add some local touches and submit them to newspapers in your area.

My thoughts on *Samuelson Sez.*

Get Me My Blood Pressure Medicine! - May 6, 2010

Please, somebody, bring me my blood pressure medicine and doctor, will you increase the dosage? I need it! Because hardly a day goes by that I don't receive an e-mail, a letter or a newspaper clipping with yet another attack on America's farmers and how they produce our food.

The latest comes from a listener in Illinois; it's a newspaper clipping from the April 18th *Peoria Journal Star*, from the Fun, Comics and Puzzles page, aimed primarily at a younger audience. One of the items on that page is an artist rendering of a Holstein cow eating grass in a pasture. It is accompanied by these words:

"Our cow is how we like to think of a dairy cow, peacefully munching grass in a meadow. But that is not how most dairy cows are raised. Most cows have a number and not a name. Almost all dairy cows are raised in factories where they are fed surplus corn and soy, not the grass that is a cow's natural food. A cow is built to digest grasses with a stomach divided into four parts; eating corn and soybeans and being crowded in a factory makes problems for cows."

The article goes on to say, "Factory cows get more diseases so they are medicated constantly. Their food includes drugs like antibiotics and hormones. They also get gas which can hurt their four-part stomachs. Factory cows live from three to four years before they die. Cows that eat grasses in meadows live and produce milk for up to 20 years. They are ones the farmers name and don't number."

Don't newspapers have "fact-checkers" anymore? The answer to that question is obvious, so let me provide some corrections. First of all, dairy cows do not live in a factory. The definition of "factory" is an assembly line with workers on the line performing specific tasks. Dairy cows are housed and milked in a dairy barn and milking parlor. Secondly, cows living in that environment live, producing milk and calves, much longer than the three to four years stated in the story. Finally, cows like those on our Wisconsin dairy farm who spent summers on grass and were named, not numbered, never made it to 20 years and were usually on their way to market in 10-12 years.

This is another reason why we need to financially support Agriculture In The Classroom and every other program that can reach an audience and respond to articles like this one that will be believed by many young people. We simply can't afford to let stories like this go unchallenged and this is why we, as individuals and organizations, need to be involved every day in telling the true story of American agriculture and reminding people... No

Jim Lucas began working at WGN in 1970 and remains a fixture at our remote broadcasts, setting up the equipment and getting us on the air.

Farms, No Food. Now, please excuse me while I go take my blood pressure medicine... My thoughts on *Samuelson Sez.*

It's the Right Thing To Do - June 10, 2010

All of us in agriculture know that abusing livestock is wrong; just like we know robbing a bank or shooting someone is wrong. But, unfortunately, abuse happens. In the case of livestock or poultry abuse, it can quickly make its way to a television screen because members of animal rights groups love to get a job on a livestock or poultry farm and then shoot undercover video until they catch someone abusing a bird or an animal. Then it gets over-the-top exposure on YouTube, network news and cable news.

I have been involved in television since 1956 and I currently produce and co-host a weekly television show. After all these years there is an element in TV that really bothers me. I can fill the screen with an incident and create the illusion that's all that's happening. Let's take the situation of an individual case of animal abuse caught on camera. Fill the TV screen with that scene and you create the impression that animal abuse happens on every livestock and poultry farm in America. We know that is not the case and that's why we must find those rare instances and stop them before they make TV news.

One of the arguments we use to refute charges of widespread animal abuse is "We treat animals well because it is good business sense." In a recent issue of *Hoard's Dairyman* magazine, an editorial entitled "It's the Right Thing to Do" suggests a different approach that I want to share with you because to me, it makes a lot of sense.

Quoting the editorial,"There's an animal care message that all of us in our industry must start putting first and foremost. It simply is this: WE PROVIDE GOOD CARE FOR OUR ANIMALS BECAUSE IT IS THE RIGHT THING TO DO. We all know that providing good care for our animals does indeed make good business sense. But we have to understand that consumers may not feel reassured about the care our animals get when the message they receive is based on economics. With few exceptions, they could care less about the margins at your farms. And, for that matter, they may not really care what the 'science' of animal care says."

The editorial concludes, "We must communicate to everyone who will listen that we provide good care for our cattle because it is the right thing to do. That's the way we should run our farms. That's the message consumers need to get, loud and clear." I fully agree with *Hoard's Dairyman*. Let's do it

Governor Jim Edgar and Bob Collins at the dedication of the Junior Livestock Building at the Illinois State Fair in 1997.

with all of our agricultural practices because "It's the Right Thing to Do."

My thoughts on *Samuelson Sez.*

Ag Challenges Ahead in 2011 - December 26, 2010

No one ever said farming and ranching would be easy. Producers of food, fiber and energy face challenges everyday caused by uncertain weather and volatile prices. I, for one, am grateful that so many of you out there are willing to do what you do to put an abundance of safe, nutritious food on my plate.

But I'm sure that many times you wonder why there seem to be more challenges every year and I'm afraid that will be the case in 2011. As I look ahead, I see an increase in attacks on producers and the methods you use to increase production. Radical environmental and animal rights groups will continue their campaigns to eliminate animal agriculture and stop the use of biotechnology on farms and ranches.

Then there is Washington, DC, with members of Congress and government agencies who will attempt to pass legislation and write new regulations that will change the way you farm, complicate your life and require completing stacks of new forms. In most cases, the rules and legislation will be written by people who have never set foot on a farm or ranch, much less had to make a living producing food.

It will require more time on your part to keep abreast of the threats, and work as individuals and through your organizations to make sure agriculture is well represented and heard. Here are some of the issues that concern me:

- Global climate change policy that in its current form will put U.S. producers at a major competitive disadvantage in the world market.

- Environmental Protection Agency rules dealing with wetlands, clean water and air and regulating dust control on your property.

- Food safety rules that would make it difficult for farmer's markets, roadside produce stands, county and state fair food vendors and small meat processors to do business.

- And of course, work throughout the year writing a new farm bill due in 2012. Rest assured, a lot of groups not directly involved in production agriculture will want to help write that legislation.

Add to all this, lawsuits filed by food processor and retailer organizations, as well as auto makers, to stop the increased use of ethanol and alternative fuels; court rulings against agricultural technology and, oh yes, the "unwanted horse" problem and we, indeed, have a plate full for 2011.

Let's be alert, involved and united and make sure the world's most productive agriculture continues to grow.

My thoughts on *Samuelson Sez*

Get Your Windshield Education - June 18, 2011

I have a term I use this time of year when I'm in conversation with my city friends. I suggest that they get a windshield education this summer. So, what is a "windshield education," you ask? Let me try to explain.

We are into the summer vacation season. Many families will be leaving homes in cities and suburbs to travel by planes, trains and automobiles to vacation destinations. I know you will be anxious to get to your vacation destination, but I suggest you make good use of the time between your home and that vacation spot by observing what is happening on both sides of the highway.

You will be traveling through some of the most productive farm land on the planet; land that is farmed and ranched by some of the most productive, efficient, caring people on the planet, the families of farmers and ranchers across the United States.

It is a good opportunity to understand and to help your children to understand that food does not originate on the shelves of the supermarket. It takes a lot of hard work by these people to get it there, and the challenges they face sometime seem insurmountable. This spring has been a prime example with dry weather in wheat country, flooding in corn and soybean country, tornadoes destroying farmsteads and spreading debris in farm fields in many parts of the country. A challenging time, indeed. Despite all of the challenges our producers face, you know that when you walk in the door of your super-market, there will be an ample supply of food there.

So, take the time to learn and observe the cattle and hog farms, the dairy farms, the corn, wheat and soybean farms, the vegetable and fruit farms and orchards and remember that 95% of those farms are owned and operated by farm families.

Finally, let me answer one question I know I will get from my readers and listeners because it happens every year about this time. People who travel by airplane across the country to the West Coast look down on farms in the Plains states and see hundreds of circles in the fields. The question comes by phone, letter and e-mail... "Why do farmers farm in circles"? I explain they are not made by space aliens, but are made by central pivot irrigation systems that travel in a circle around the well in the center of the field to provide the necessary water to grow the crop. I'm glad the question is asked because it makes for another educational opportunity. Do take the time to get a "windshield education" this summer.

My thoughts on *Samuelson Sez.*

DEFENDING THE MEAT INDUSTRY

Let's Tax the Real Polluters! - December 11, 2008

It is five years ago this month that America's cattle producers received an unwanted Christmas gift, the discovery of "the cow that killed Christmas." The first case of Mad-Cow Disease was discovered here in the U.S. in the State of Washington.

Now, five years later, cattle producers and other producers of livestock and poultry could get another unwanted gift in the form of a tax on animals; and I have to choose my language carefully here, but I'm talking about animals that belch and pass gas. The bureaucrats in Washington are now talking about a Federal proposal to charge fees for "air-polluting" animals. The proposal that is being floated would require farms or ranches with more than 25 dairy cows, 50 beef cattle or 200 hogs to pay an annual fee of about $175 for each dairy cow, $87.50 for each beef animal and $20 for each hog, and let me repeat, it would be an annual payment.

Ken Hamilton, the Executive Vice President of Wyoming's Farm Bureau, said the

One of the perks of my job is judging food contests. In this case, it's a beef cook-off.

Sharing a laugh with Governor George Ryan and Illinois Director of Agriculture Joe Hampton at a party the Governor hosted at the Illinois State Fair in honor of my 40th year at WGN Radio.

fee would cost owners of a modest-size cattle ranch $30 to $40 thousand a year, and would bankrupt them. Ron Sparks, the Agriculture Commissioner in the State of Alabama put it very well when he said, "This is one of the most ridiculous things the Federal Government has tried to do." He said it will mean that, "We will let other countries put food on our tables, like they are putting gas in our cars, and other countries don't have the health standards we have."

Of course there are those who are applauding the EPA proposal and you can probably guess who is offering strong support... PETA. A spokesman in Washington for the People for The Ethical Treatment of Animals, said and I quote, "It makes sense if you are looking for ways to cut down on meat consumption and recover environmental losses. We certainly support making factory farms pay their fair share."

But Perry Mobley, Director of the Alabama Farmers Federation Beef Division said "Who comes up with this kind of stuff? Somebody who obviously doesn't appreciate what it takes to put food on our tables." And I am in total agreement with Ron Sparks of Alabama. This is a ridiculous idea!

Producers, be ready to fight this one and maybe part of our counter-attack should include the suggestion that we tax the bureaucrats for all the gas they pass when they dream up ideas like this one!

My thoughts on *Samuelson Sez.*

Listen up, Cattlemen... They're At It Again! - May 9, 2009

While pork producers have been on the receiving end of negative news the past couple of weeks, I certainly don't want people in the cattle industry to feel ignored or neglected. So here comes a dose of negative news for you from Wildlife Guardians, an organization that just published a report entitled "Western Wildlife Under Hoof." For the past year this organization has used satellite mapping and federal records to match wildlife habitat and U.S. grazing allotments across more than 260 million acres.

And cattlemen, pay attention, because this is what the report had to say about you and your industry: "Cattle and sheep trample vegetation, damage soil, spread invasive weeds, spoil water and deprive native wildlife of forage. Continued grazing in ever-shrinking habitat hampers the recovery of fish and wildlife and in some cases threatens them with extinction." Did you notice? No mention of the impact of man's intrusion with housing developments and shopping malls into what was wildlife habitat. Mark Salvo, of the Wildlife Guardians said — and you'll love this one: "Livestock have done more damage to the earth than the chainsaw and the bulldozer. Combined."

Quick to respond, Jeff Eisenberg, Director of Federal Lands for the National Cattlemen's Beef Association said, "There are a number of environmental groups that have decided the best way to spend the time and money of their funders is to eliminate the families and communities that have made the West what it is today. These groups don't deserve a dignified response."

You know where I stand on this issue. I totally agree with Jeff! If, indeed, Wildlife Guardians were successful in putting a halt to grazing on public lands, it would affect 15,779 ranchers covering 128 million acres of public land that these ranchers and their families have improved for wildlife habitat.

You know, I really wonder how this planet ever made it this far, because for centuries in North America there were no people around to fight forest fires, no one to worry about millions of buffalo roaming the range and burping harmful emissions into the atmosphere, no one to worry about spotted owls and other endangered species. I wonder how many species arrived and became extinct before humans made their first appearance. But here we go again with another group and another attack on the livestock industry. Let's not take it lying down.

My thoughts on *Samuelson Sez.*

Tell Us How You Feel About GIPSA - September 4, 2010

The USDA has more than its share of acronyms, but the one that is getting a great deal of attention now is GIPSA. Some of you are asking, what is it, what do the letters mean? It is the 89-year-old Grain Inspection, Packers and Stockyards Administration that oversees the marketing of grain and livestock in the United States. The Department of Agriculture says it is time to make some changes, to level the playing field for livestock and poultry producers.

It's a challenge to know exactly what all of the changes are, but here are the top five according to those who have studied the USDA proposal:

#1 - Force meat plants and buyers of livestock and poultry to maintain written records that justify variations in prices offered to producers.

#2 - Make it a violation to offer better prices to big producers who can provide larger volumes of livestock than to smaller producers who, collectively, can provide the same number and quality of livestock.

#3 - Define unfair and anti-competitive practices so that violators can be punished in court. An example would be using inaccurate scales to weigh poultry.

#4 - Require livestock buyers to work with only one meat packer.

#5 - Prevent a meat packer from buying livestock from another packer. USDA says this would prevent collusion and open the market to more participants.

Here we go again with another very emotional debate, nearly as emotional as the suggestion of a national animal ID program a few years ago. This one seems to put small producers against large producers. R-CALF, a trade group representing smaller producers, plus a number of other grassroots producer groups, say the possible changes would help restore marketplace competition.

On the other hand, Don Close, Marketing Director for the Texas Cattle Feeders Association said it kills the incentive for the producer who strives to have the superior genetics and management and makes the investment to consistently produce top-quality beef that attracts a higher price.

As I said, the arguments are angry and emotional suggesting to me there is probably a little "right" on both sides. I would be interested in hearing from you. Should GIPSA be changed? Should government become more involved in livestock marketing and pricing? Send me your thoughts and reasons and I'll share them with my national audience on upcoming editions of *Samuelson Sez.* E-mail me at agbizweek.com.

My thoughts on *Samuelson Sez.*

DAIRY FARMING

Science battles Greed - January 27, 2008

A few days ago I received an e-mail from an angry, frustrated dairy farmer in Illinois. After reading the e-mail, I quickly shared his anger and frustration. Let me quote part of it. "Dear Orion, We had a real bad news day today. Our field man called to say Dean Foods at our local processing plant will no long buy milk from cows supplemented with rBST as of February 1st. Walmart contacted Dean Foods to say it would no longer buy rBST milk and if Dean's wouldn't go along, it would buy its milk elsewhere. Dean's pleaded that it needed more time for their producers to use up their current supply of rBST, and Walmart's response: 'That's not our problem'." End of quote.

Here is another prime example of emotion or maybe financial greed over-riding sound science. First of all, there is no difference in the milk from a cow that produces BST naturally or another cow that has been supplemented with additional BST to increase milk production. There is no scientific equipment that can detect any difference. This is not sound science and certainly not sound reasoning.

For whatever reasons, among them perhaps financial, dairy processors and dairy marketers at the retail level label their milk hormone-free, and then raise the price. If you check the price of milk that is labeled "free of rBST" you will find the price is considerably higher than regular milk. In the second quarter of 2007, the Farm Bureau Market Basket Survey showed the price of a half-gallon of whole milk to be $2.22, the price of a half-gallon of milk labeled rBST-free was $3.01, a 36% premium for milk that is no different. I doubt if any of that 79-cent premium reached the producers who signed the "no rBST affidavit" to satisfy the demands of the processor.

Golly, I share this frustration. My bet is people who make these decisions at Walmart have never milked a cow, have never been involved in food production and give little thought to what stopping the use of technology in food production will do to hungry people around the world. It will literally take food out of their mouths.

This is a wrong decision by the dairy processing industry, because if you can't test it, how can you label it? Agriculture must come together on this one, and spend time every day educating people, both in the industry, as well as consumers that science, not emotion or financial gain, must be the benchmark when evaluating the use of technology in food production.

My thoughts on *Samuelson Sez.*

Thank God for Farmers and Ranchers - June 11, 2009

We all know that life is full of change and challenge on a daily basis; but as I look at America's farmers and ranchers, I think they get more than their fair share. That is why every day I say "Thank you, God, for America's farmers and ranchers; the people who put food on my table, clothes on my back, a roof over my head and energy in my tank."

For those of you not involved in production agriculture, let me list just a few of the challenges faced by farmers and ranchers in 2009. A wet spring that has left a few million acres still unplanted or in need of re-planting; pork producers who have been hammered economically the past year, now getting a double hammer from the H1N1 flu virus that has cut their income

My love of cheese, which seems to have started at birth, led me to buy several winning cheeses at the Illinois State Fair, including this Grand Champion Wheel of Bleu crafted by Hans Michael of Navoo Cheese Company in 1990.

even more; and dairy producers who are in financial pain dealing with milk prices that are $6 to $7 per hundredweight below the cost of production.

The pain becomes very real when you share it with me via e-mail as was the case last week when I heard from Charles in Connecticut who wrote "Dear Orion, I am hoping you can share some facts about the New England dairy industry. In Connecticut it costs $2.00 to produce a gallon of milk. The price of milk which is controlled by the Federal government is only $1 per gallon. This means that Connecticut dairy farmers are losing $1 per gallon of milk produced and with a milk price set as low as it is, small states such as Connecticut are losing their dairy industry. According to the Connecticut Department of Agriculture, the state has a $1.1 billion dairy industry; the state currently has 151 dairy farms after 12 shut down last year. If people do not contact their legislators and ask them to raise the milk price, the dairy industry around here will disappear. Thank you for taking the time to read this and I hope your listeners and readers from all over the country will help out our farming industry here in Connecticut."

Adding to my concern is the fact that, after sharing it with my audience, there is little I can do to help Charles and the rest of you faced with similar challenges. I can say to Charles that he is not alone, there are many other producers in the same boat, but that is little consolation. I can also say something you have heard me say many times... prices never go the same way forever, there is always a correction, but for some of you it may not come in time.

So I come back to the thought that started this column: thank God for all of you in production agriculture for facing the challenges of weather, markets, trade policies and politicians every day, as well as the criticism of consumers when food prices go up. And the next time you are asked to offer the invocation at a banquet, besides asking God to "bless the hands that prepared this food," will you also ask to "bless the hands that produced it"?

My thoughts on *Samuelson Sez.*

Could We Survive the Dairy Farmer's Challenges? - December 4, 2009

A few days ago I received a letter from Dale Anderson. He's a dairy farmer in Randolph, New York who milks 80 to 90 cows on his dairy farm. He has heard me talking about the challenge facing hog and dairy producers and he shared with me a presentation he made at the New York State Senate hearing on "Saving the Dairy Industry in New York" last month. The hearing was chaired by Senator Cathy Young and three other state senators. He was one of four dairy farmers asked to make a presentation and he described the plight of dairy farmers from first-hand experience.

This is some of what he said: "As long as the store shelves are well stocked at a low price to the consumer, and people are well-fed, most people, politicians included, aren't very concerned with the plight of the dairy farmer.

"In the month of July 2008 my milk receipts were in excess of $22,000. In the same month of 2009, they were less than $8000. How many of you could stand a $14,000/month cut in pay? I can't either!! I have virtually hit the wall as far as being able to run a profitable dairy operation. I am on a cash-on-delivery basis with my feed supplier. I have been forced to cut back considerably on the amount of protein grain I feed my cows and have entirely quit feeding corn meal.

"Several times I have had to pay my electric bill with post-dated checks to keep the power on to milk my cows. For the last four months my wife has had to cook with a hot plate and no oven as I owe over $1,700 for propane gas and they will not deliver without full cash payment. When Senator Young called me to invite me to speak at this forum, I was glad I had just paid the disconnect notice on my telephone bill or we wouldn't even be able to have this conversation!

"Some banks are no longer making farm loans. I still owe part or all of my bills for last spring's planting. I have also not paid for the cost of harvesting my corn silage. At the present time, my checking account is in an overdrawn balance. My wife has maxed out her credit cards trying to keep the farm afloat."

This is a heart-breaking story told by New York dairyman Dale Anderson, and making it even worse, is the fact there are many others who are living a similar story. Milk futures are beginning to show a price recovery and a return to break-even by next summer. But that may not be soon enough for many of our friends who milk cows. Think kindly of those people the next time you enjoy a pizza, an ice cream sundae or drink a cold glass of milk.

My thoughts on *Samuelson Sez.*

What We Have Learned from 2009 - December 19, 2009

As we approach the end of 2009, with a 35% drop in net farm income and major price challenges to hog and dairy producers, I guess most of us would say, "Good riddance, let's look forward to a better 2010." Well, my topic this week comes from a headline on the cover of the December issue of *Dairy Herd Management* magazine. The headline, "What we have learned from 2009."

The article quotes six dairy farmers around the country, talking about the lessons they learned in a very challenging year that is finally coming to an end. Let me share some of what they had to say.

Gale Moser, Moser Dairy, Whitney, Idaho: "Don't spend it before you make it."

Brad and Brice Scott, Scott Brothers Dairy, San Jacinto, California: "Be thankful for what you have."

Liz Doornik of Jon-De Farms, Baldwin, Wisconsin: "Be open and honest."

Ted Boersma and T. J. Curtis of Forget-Me-Not Farm, Cimarron, Kansas: "Don't forget what's really important."

James Davis and Ron and Lydia Lewis of TA-RO-LEE Holsteins, Chowchilla, California: "Out of every negative comes a positive."

And finally, the sixth response to the question: Ryan Anglin, Triple A Farms in Bentonville, Arkansas said, "Don't forget your family." He said, "Family is always important, but in times like these, it becomes even more important. Family is the only support you really have. But employee loyalty has also played a role in our business surviving this year. Our employees have stuck by us. They've gone without raises and have been truly an asset getting through this year."

Then looking to the future he said, "I will look at forward-contracting my milk and grain to eliminate some of the ups and downs of the price roller coaster."

What I like most about this story is the positive tone. Those six dairy farmers have gone through the toughest year the industry has experienced in 50 years and yet, I didn't hear any of them say, "feel sorry for me." Instead, I heard them say, "We learned something and hopefully it will make us better business people in 2010." And none of them lost sight of what is truly important... family and a positive mindset. Now, my wish for them and all producers in 2010 is prices that will be higher than the cost of production.

My thoughts on *Samuelson Sez.*

FOOD AND FARM PRICES DON'T GO THE SAME WAY FOREVER

Enjoy the Moment - September 9, 2007

Notice the prices for hogs and cattle on the wall behind me in this 1962 photo from Top O' the Morning: $16.75 cwt for hogs, $29.75 for cattle.

In the last week of August, the U.S. Department of Agriculture issued its revised estimate on U.S. net farm income in 2007. The numbers were impressive, encouraging and long-overdue. Let's look at the numbers before I state my challenge to you.

First of all, the report said net farm income in 2007 will be nearly 48% greater than a year earlier at a record $87.1 billion. That prompted Chris Hurt, long-time economist at Purdue University, to say, "This is a great time to be a farmer. Farming may be the healthiest sector of the economy." In light of the challenges in the auto industry, housing and stock markets, I would agree.

Let's compare the current numbers with a year ago. Total value of crop production in 2007 is estimated to be up 14%; price of corn up 62%; price of wheat up 46%, to all-time record highs; price of soybeans up 36%. The value of livestock and poultry production for this year is up 18% from a year ago, because of production cuts by poultry and egg producers; falling global dairy supplies; U.S. cattle inventory at a 40-year low, pushing cattle futures prices above the $100 mark at the Chicago Mercantile Exchange. So indeed, these are impressive numbers.

But now let me quickly say, I realize that not every producer in America benefits from these higher prices. And after I carried this story on my daily radio show, I received several e-mails. Some of them said, "Don't tell people we are making money. Don't tell them times are better because then they will say there is no longer a need for government farm programs." If times are better, why not say so? If city union workers get a contract with higher wages, they shout it to the world and people cheer them. Perhaps in agriculture we have spent too much time talking "poor mouth" and built a mind-set in non-farmers that we should never make money.

Then, of course, I heard from others saying, "Yes, Orion, our prices are higher, but our production costs are way up, too." Yes, there's no argument there, production costs are higher. But if you have done a good job of marketing (and that means not holding last year's corn, because you knew the price was going to $5.00), I think the balance sheet will show a profit in 2007. And finally this e-mail, "Prices may be good now, but I'm worried about next year."

Having grown up on a farm, I know farmers always have to worry about something, but may I make a suggestion? At least take a moment to enjoy today's prices and admit that this year is a pretty good time to be a farmer. Then, before you start worrying about next year, look around you and see what brought these better prices and work to keep those conditions in place to ensure a profitable agriculture for next year and beyond. Enjoy the moment!!

My thoughts on *Samuelson Sez*

Expectations and Final Results - August 16, 2008

The best part of this job is traveling the country to agricultural events and visiting with you on the other side of this column.

A few days ago I made my annual trip to the Illinois State Fair in Springfield and had the opportunity to visit with quite a few producers. It was a good day to be there; it was the day of the August Crop Report. Over the years I have learned that while the numbers in those monthly crop reports change, the reactions don't.

At the fair, I talked to farmers who said, "Yeah, the crop has really come on and I think those numbers are right on." Then I hear from others who say, "No way will the crop be that big. Too many late-planted fields in my area and what if we get an early frost?" I always have to remind producers that you can't judge the total national crop based on what you see in your neighborhood. Secondly, the numbers in the monthly reports take into ac-

With Warren LeBeck, president of the Chicago Board of Trade in the mid-'70s.

count weather and growing conditions at the 1st of the month and assume normal growing weather from that point forward.

I also heard this comment that came as no surprise after a market sell-off. A corn farmer said, "Samuelson, what did you do to the corn price at the Chicago Board of Trade? For heaven's sake, it dropped like a rock without any reason." I looked at him and said, "I'll bet you didn't sell any at $7 a bushel." With a sheepish grin he said, "No, I thought the price would go higher." That is not a marketing plan, that is Las Vegas gambling and once again, let me remind you that prices never go the same way forever, there is always a correction.

In the grain trade, we always get private estimates on the numbers ahead of the report and we know who makes those estimates, or "guesstimates" as I like to call them. But I want to turn to Wall Street for a moment and share something that really bothers me there. On my daily radio shows in Chicago, I report agricultural markets, as well as Wall Street. I am really concerned by the unidentified analysts who set "expectations" of what earnings reports should be, and if the report shows a profit, but doesn't meet the "expectation," the stock price plummets.

Case in point: a few days ago John Deere issued its quarterly earnings report. That report showed net profit increased 7% over the same quarter a year ago, a pretty good number of $1.32 a share. But those unknown analysts expected $1.36 a share, four cents more, and when they didn't get it, they drove the price of Deere stock down 12% that day. I don't own any Deere stock, but to me that reaction to a profitable earnings report makes no sense. I still don't know who these unknowns are, but I wish I could find them and sell their stock down 100% and put them out of business.

My thoughts on *Samuelson Sez.*

Patience, not Panic is the Key - August 12, 2011

A reader fired off an e-mail to me last week, using some language that I can't share in this column, but his question basically was, "What is going on in the marketplace?" I fired off a one-word response. "Volatility!"

Volatility has always been in the marketplace, whether it is Wall Street or in commodity trading in agriculture, and yes, it is increasing. There is no question in my mind that as we have moved from strictly trading in the pits by human beings, to now trading electronically with computers around the world directing much of the trading, volatility has increased. Add to that several 24-hour TV business news channels with dozens of business "experts" expressing their opinions on why the market is up or down sharply and it all leads to confusion which leads to uncertainty which leads to fear and pushes people to make decisions and take actions they ought not take.

That was driven home to me in several e-mails that read like this: "What is happening to my retirement fund, my 401k? I am so frightened by this 600-point drop in the Dow that I'm selling my stock and getting out of the market until things calm down." That is the worst thing an investor can do, panic sell, which over the years I've seen happen many times on Wall Street as well as in grain and livestock futures markets. We certainly saw it this month when after the 600-point drop in the Dow, the market recovered more than 400 points the next day, leaving those who sold Monday with a large (and real) loss in their retirement fund.

That same week we saw the "limit up" move in corn futures following the August Crop Report, and I talked to some traders who bought at the top because they knew the corn price would go higher. Perhaps it will, eventually, but the next day the price went down as other traders thought that we have a long way to go before the crop is in the bin and a lot can happen between now and the end of harvest. That makes for volatility and it is not going away.

Some things to remember: Number one, try to stay calm, don't sell in a panic market, don't join the "leming parade" and follow the leader over the cliff. Number two, and you have heard me say this many times, "Prices never go the same way forever. There is always a price correction." That is the one lesson I have learned in my half-century of covering markets and is the only prediction I can make without fear of it being wrong.

My thoughts on *Samuelson Sez.*

ETHANOL

I Agree With Farmer Rapp - May 2, 2008

I can attest to the fact that the "food vs. fuel" controversy is escalating. Some of it is getting downright nasty. I have received some unkind e-mails recently, not just from city people who are against the idea of using corn for ethanol, but also from farmers who feel the same way. As a matter of fact, one e-mailer who identified himself only as a livestock farmer, told me I should "engage my mind before I engage my mouth" and he was extremely critical of my support of corn and ethanol as part of our energy solution.

So, I will use someone else's mind today and quote a letter to the

Introducing Jim Ryan, Illinois Attorney General from 1995-2003, at the Illinois State Fair.

editor written by Jim Rapp, a northern Illinois corn farmer. I can't quote the entire letter because of space limitations, but here are some excerpts. In the letter to the editor of his local newspaper, Jim wrote, "The rapid growth of the ethanol industry tells you all you need to know about its performance and viability. It burns cleaner, thus fighting air pollution, ask the American Lung Association; it is made here at home, providing jobs, just ask the 230,000 people already employed as a result of ethanol plant construction, operation and related support services; it provides more dollars into our local communities, just ask local business owners about the increased dollars spent at their stores."

Jim goes on to say,"Consumers who have seen gasoline prices climb more than 40% over the last four months want relief and alternatives. We owe it to them and ourselves to explore all of our engine technology and renewable fuel options if we want a fully functional and strong U.S.A., which is less reliant on foreign fuel sources. Right now ethanol is leading the assault to solve this problem and, contrary to opponents of this petroleum alternative, ethanol is a terrific fuel with a proven 67% net energy gain."

Farmer Rapp concludes,"Do not believe everything you hear about

Governor James "Big Jim" Thompson was Ilinois' longest serving governor, from 1977-1991. He was the first candidate for governor to receive over 3,000,000 votes and was a big supporter of Illinois agriculture.

corn's role in food prices. The number one factor driving food price increases is, you guessed it, transportation. Fuel prices are sucking our economy into a black hole. When you add that to recent rice and wheat crop failures overseas, speculative investors shifting to agricultural commodities in our weak economy and growing populations with increased buying power in China and India, you quickly see corn and ethanol barely make a ripple in the pond of a much bigger issue. Even at today's food prices, farmers get no more than 20 cents of each dollar spent on food. A box of cereal that contained 2.5 cents worth of corn 18 months ago now contains a nickel's worth of corn. I think you get the idea, agricultural commodities are not a major factor in the price of food."

All can say is Jim Rapp said it very well and I fully agree with him. That's why I have shared his thoughts on *Samuelson Sez.*

Statistics... Fact vs. Fiction - June 8, 2008

The statistic has been around since the 1980s and, over the decades, supporters of ethanol have had to deal with it over and over again and haven't been able to put it away. As recently as a few days ago, on the Fast Money Show on CNBC, it reared its ugly and false head once again. The statistic? That it takes 1,700 gallons of water to produce a gallon of ethanol.

Not true! But it was born in the 1980s when Dr. David Pimentel, an entomologist at Cornell University launched a campaign to refute claims of the nation's corn growers that ethanol could be an alternative fuel to help cut our reliance on foreign oil. For some reason, he didn't like ethanol and to support his argument, he included the total amount of water needed to grow the corn crop in addition to the water needed to process the corn into ethanol to reach his 1,700 gallon claim. Never mind the fact that as little as 4% of the corn used for ethanol production in this country requires any irrigation and that God-given rainfall provides the moisture for the other 96% of the crop.

So, how much water does it take to process ethanol? It takes four gallons of water to process one gallon of ethanol. That's less water than it takes to flush the average toilet in this country, so please, let's put this false statistic out to pasture forever, so we don't have to deal with it again. While I'm at it, let me add one more number to the mix. It does take 1,851 gallons of water to refine a barrel of crude oil. That's according to Argonne National Laboratory which also says water use in dry mill ethanol plants dropped more than 26% between 2000 and 2006.

On another subject, but still related, I fully expected speakers and attendees at the Global Food Summit in Rome last week to spend most of their program time bashing the bio-fuels industry, blaming it for world hunger and starvation, higher world food prices and all the other ills of the world. So it came as a pleasant surprise that while it was mentioned, speakers focused more on trade policies that limited movement of food and agricultural products between countries. They were especially critical of countries that embargoed shipments of food to other countries to ensure their own citizens had enough food; and some speakers had very harsh words for the oil-rich countries who donate little or no money to fight world hunger.

There were some very positive suggestions on steps to take to alleviate world hunger and I especially liked the idea of a new Green Revolution to equal the success of the first Green Revolution of the 1960s led by Nobel Peace Prize winner, Dr. Norman Borlaug, a program that literally saved millions of people from starvation. We did it before, we can do it again!

My thoughts on *Samuelson Sez.*

Food and Fuel Production Can Co-Exist - August 8, 2008

Judging from the e-mails and letters that come across my desk, the "Food vs. Fuel" controversy simply will not go away. I hear from folks who are strongly opposed to ethanol and soy-diesel; opposed because they say it's immoral to take food from the mouths of hungry people to turn into energy for our gas-guzzling SUVs. They blame ethanol for sharply higher food prices and world hunger. Then I hear from the supporters of ethanol, who say we can do both, we can produce food and renewable energy on America's farms, something we need to do, to cut our dependence on foreign oil.

Throughout the debate, I have said the primary reason for sharply higher food prices is not the production of ethanol or the ethanol subsidy; it is the sharply higher price for crude oil that is used in every phase of agricultural production, food processing and transportation.

Now, a report from the Farm Foundation taken from a symposium in Washington, DC in July entitled "What's Driving Food Prices?" pretty much says the same thing. I'd like to quote one paragraph from the 78-page report... "Most of the corn price increase is due to the higher oil price, not the subsidy. With no subsidy or mandate, corn moves from $1.71 a bushel at $40 oil to $5.26 a bushel at $120 oil. With the subsidy, corn moves from $2.26 at $40 oil to $6.33 a bushel at $120 oil. Put in round numbers, when crude went from $40 to $120, corn went from $2 to $6, a tripling of both prices. About $1 of the corn price increase was due to the subsidy and $3 to the higher crude price. As the price of oil has increased, corn-based ethanol is demanded to substitute for gasoline. At high oil prices, this would happen, with or without the subsidy." Finally, someone puts the blame where it belongs, on high oil prices.

We need to remember something else about agricultural commodity prices, they do come down; to those food processors wringing their hands and increasing their prices (while reporting higher quarterly earnings) because the cost of the raw commodity moved higher, take note. Through the first week of August, the price of wheat had dropped 50% since February, the price of corn had dropped 30% since June and the soybean price had dropped 30% since early July. We have even seen a 20% drop in the price of oil! But we certainly haven't seen a similar drop in the price of bread or other food products.

It is also worth noting that feed costs for cattle and hog producers have declined while hog prices have strengthened and cattle prices have remained strong. So producers, let's get together and work together to respond to those critics and assure them we can produce both food and energy on the farms and ranches of the U.S.

My thoughts on *Samuelson Sez.*

Here Are the Facts! - January 20, 2011

While the Internet makes it easy to find a great deal of useful information; it also makes it easy to find a great deal of misinformation or in some cases, outright false information. Anyone can put anything on the Internet and someone will read it and believe it without checking other sources to confirm the information.

Judging from my recent e-mails, there are two agricultural issues coming under attack by groups or individuals that have little or no basis in fact. First, the size of the USDA farm budget and why it should be eliminated to help solve our national debt crisis, and secondly, stop making ethanol from corn, the "food vs. fuel" argument. So let me steal a line used many times by Sgt. Joe Friday on Dragnet. "Here are the facts, ma'am, just the facts."

First, let's put budgets in perspective. The federal budget is three trillion, eight hundred thirty billion dollars. This is what it looks like in numbers: \$3,830,000,000,000. The U.S. Department of Agriculture budget is one hundred forty nine billion dollars or \$149,000,000,000. Yes, that is a lot of money, but it is less than 1% of the total federal budget. Where does the money go? Seventy percent of the USDA budget goes to Nutrition Assistance which includes the old food stamp program, now called SNAP, and other nutrition programs; 17% goes to farm and commodity programs to actually benefit the nation's farmers and ranchers; only 7% goes to conservation and forestry; the other 6% goes to research, food safety, and other programs administered by the USDA. Just remember, the USDA budget accounts for less than 1% of the federal budget; eliminating it wouldn't even make a blip on the federal budget radar screen.

Then, the argument over "food vs. fuel" is heating up again and several of you have told me about a blog that states it is absolutely sinful to take ethanol out of a bushel of corn and throw the rest away. First, let me say that simply is not true, we get a lot more than fuel and that's why I call it "food and fuel." Again, here are the facts.

From one bushel of corn, we get 2.8 gallons of ethanol, but as my friend, the late Paul Harvey would say, "This is the rest of the story." We get 18 pounds of feed for livestock and poultry; we get corn oil that is used in many foods for human consumption and we get other products like CO2 that makes bubbles in your soft drinks and dry ice. No, we do not throw away the rest of the bushel after we make ethanol and that's why it's "food and fuel," not "food vs. fuel."

Thanks for bringing these stories to my attention and you may want to adopt my policy on Internet stories, check at least three other sources to verify the information. After all, it is very important to get "the facts, ma'am, just the facts."

My thoughts on *Samuelson Sez.*

If We Cut Subsidy, Do it Fairly! - April 6, 2011

With corn prices hitting record highs this year and global food prices hitting record highs, the makers of ethanol again are under fire from all sides, including cattlemen, dairy farmers, poultry producers, hog producers and those who say "we ought not take food from humans to put into the gas tank."

Let me, first of all, remind those people that after we take the ethanol from a bushel of corn we still get livestock feed in Distilled Dried Grains and human food in the form of corn oil coming from that same bushel.

But on *Samuelson Sez* over a year ago, I went on record saying it's time to end the subsidy for ethanol, IF, and it's a big IF as far as I'm concerned, IF we also take away subsidies from the producers and processors of oil and gas, nuclear and coal, and all of the other energy sources, but oil and gas particularly. Try to research the amount of government subsidy that producers of oil and gas get from the U.S. Treasury. It is a tough number to find. I saw estimates ranging anywhere from $35 billion to $100 billion a year, but the number that came up more often was a $45 billion taxpayer subsidy to the petroleum industry to provide us with energy that's now costing us well over $4.00 a gallon at the pump.

I have said the time will come when ethanol will have to stand on its own merits without that $6 billion subsidy. The producers of ethanol are aware of that. After a difficult fight, they did get a one-year extension on that subsidy. But now, bills have been introduced on Capitol Hill to do away with that legislation and end the subsidy immediately. Leaders in the industry, wanting to avoid another legislative fight, are discussing ways to moderate or change the subsidy to make it more acceptable to its critics.

The subsidy was put in place to help a new industry establish itself and ethanol is still relatively new. So why, some people ask, does the petroleum industry, well over a century old, still deserve an annual $45 billion subsidy? The rest of the energy subsidy list reads like this... $9 billion for nuclear, $8 billion for coal, $6 billion for ethanol and $6 billion for renewable fuels... wind, solar, biomass and geothermal.

So, I'll say it one more time. I am in favor of doing away with the subsidy for ethanol, IF we also cut back sharply on the subsidy for the petroleum industry.

My thoughts on *Samuelson Sez.*

Here We Go Again! - May 8, 2011

"Here we go again!... Cattlemen wanting to take money away from corn farmers and keep us from making a better living."

That was the opening line in an e-mail I received a few days ago from a Midwest corn farmer who occasionally shares with me his thoughts on various issues. He was referring to the statement made recently by Bill Donald, President of the National Cattlemen's Beef Association, supporting a Senate bill that "would end 30 years and more than $30 billion of taxpayer support for the corn-related ethanol industry and would finally level the playing field for all commodities relying on corn as a major input." If passed, the bill would end the ethanol tax credit and the tariff on imported ethanol no later that June 30th this year.

The corn farmer's e-mail ended this way: "I repeat what I told you three years ago when I was selling my corn out of the field at less than $1.50 a bushel; I don't hear any cattlemen saying thank you or feeling sorry for me. So why all the tears from cattlemen now when they are getting record high prices for their cattle? Why shouldn't corn farmers be able to enjoy record high prices, too? Cattlemen say their feed costs have gone up; what about my costs for seed, fertilizer and fuel? They are all sharply higher, too."

Both sides state their feelings well and certainly with emotion. But you know what bothers me about this exchange. Here we go again with producers shooting at each other, when all of us in the agriculture family need to be working together. There are more than enough organizations and individuals outside agriculture shooting at us and that's where we should be directing our time, energy and emotion; responding to charges from them that have little or no factual or scientific basis, but if enacted, would impose

Photo courtesy: IBA Archives, George Burns

In 2011, the Illinois Broadcasters Association asked me to introduce the newest member of the Illinois Broadcasters Hall of Fame: Oprah Winfrey. The "Big O" and the "Bigger O," I suppose. While Oprah and I both grew up on farms, we are a country mile apart on our opinions about the beef industry.

new regulations that would greatly curtail the way you produce food, fiber and energy.

While I'm talking about information that is not factual, here is another example. The Director General of the U.N. Food and Agriculture Organization recently stated,"The rising output of bio-fuels is contributing to global food shortages, consuming more than 100 million tons a year of cereal grains that would otherwise be used in food production." Once again, he is fostering the illusion that once we produce the ethanol, we throw the rest of the bushel of corn away. That is simply not true and we need to hammer home what I said in this column in January. After we get the 2.8 gallons of ethanol, we get 18 pounds of livestock and poultry feed, corn oil that is used in many human foods as well as several other products.

I hope the next time I say "here we go again," I'll be referring to a unified agriculture speaking with one voice to preserve and enhance the most efficient food production system on the planet.

My thoughts on *Samuelson Sez.*

Ethanol War Continues - July 3, 2011

During my time in Norway last month visiting with Norwegian farmers, there was one topic that never entered our conversation: ethanol. There was no discussion on corn-for-ethanol pushing food prices higher or putting livestock producers out of business because of high feed costs. There are two obvious reasons: the Norwegian climate allows very little corn production, and in Norway oil is king; it is now the third largest oil exporting country on the planet.

However, before leaving, once again I was really troubled by the war of words between corn farmers and livestock producers through their commodity groups after a 73-28 Senate vote to immediately end the ethanol subsidy and import tariffs. Even though the final vote is expected to lose, the National Cattlemen's Beef Association hailed it as an indication that Congress is ready to "level the playing field" and bring feed prices back to levels to encourage livestock production.

On the other hand, corn growers said it's their time to turn a profit and pointed out that cattle and hog prices have been holding strong during the spring and summer. They also reminded members of Congress that ethanol cuts our dependence on foreign oil and ethanol plants provide employment in rural America.

I understand the concerns of both sides.

But now, fast-forward to June 30th and the USDA report that "leveled the playing field" without any help from Congress, again proving several points: if the financial incentive is there, America's farmers will produce more; prices never go the same way forever; and "the cure for high prices is higher prices and the cure for low prices is lower prices." The second largest planted corn acreage since 1944 and a shocking 11% increase in corn stocks caused one major trading firm to lower its 6-month corn price projection from $7.80 a bushel to $5.75. That should bring a smile to livestock folks.

But, I have one more question. Livestock producers were quick to applaud the Senate for its ethanol vote, but where were they earlier this year when Congress said it would not even consider legislation to cut subsidies to oil-producing companies? Why did I hear no expression of concern about leveling the playing field between big oil and alternative energy producers? In my opinion, the price of oil has far more impact on the bottom line of all producers than the ethanol subsidy.

My thoughts on *Samuelson Sez.*

GETTING ALONG WITH NEIGHBORS

Say "Welcome" - August 20, 2006

Hardly a week goes by that I don't receive a letter or an e-mail from a farm family, terribly concerned about the fact that they have new neighbors; a family from the city that has moved into their dream country home, right next door to the farm. Suddenly the new neighbors are demanding all sorts of things. They don't like machinery noise early in the morning or late at night. They don't like the odor that is coming from the livestock farm. They don't like the residue, mud or manure, left on the roadway by farm equipment because that dirties their "newly washed car" that they drive on that same roadway. On and on it goes, the litany of concerns that ultimately, in many cases, will lead to a lawsuit. We have heard it time and again from people who are really finding themselves at their "wit's end" because they have been operating that farm for 100 years with neighbors who understand life in the country. Now suddenly, they have neighbors who have their own perception of country life that comes nowhere near matching reality.

A suggestion that I have made several times over the years, and I think it needs to be done on a county by county basis, is to form some kind of what I call a "Welcome Wagon" organization that used to exist in cities and suburbs. Its purpose would be to greet and welcome the new family to the country. I think it is important you get acquainted with these new neighbors quickly, and explain to them that life is going to be different in the country. They are going to hear noises they have not heard before; they are going to smell odors they haven't smelled before; there is no garbage pick-up every week, and yes, during thunderstorms the electricity tends to go off sometimes. There are several County Farm Bureaus in Illinois that have put together a brochure designed for city folks moving to the country. It explains all of these things to new rural residents, and it does it in an understandable way. I have found that if you make friends with people right away and explain to them what you have to do to put food on their table, they often will become your strongest supporters.

One other point: some farmers bring on the problem by being suspicious of the motives of a new city neighbor or flatly saying they have no right to be there. They do have that right and so my suggestion is when the new neighbors from the city move in, be there on the front doorstep with a chocolate cake or an apple pie and say "Welcome to the neighborhood! We're glad you're here."

My thoughts on *Samuelson Sez.*

NIMBY's, Get a Life! - November 30, 2008

I think I am a tolerant person. I try very hard to be, and I think I succeed fairly well. But I admit there are times when I lose patience. I am at that point now with NIMBY's, the "Not In My Back Yard" people. I have talked about NIMBY's before, but I continue to hear from farmers and ranchers across the country who tell me they are unable to expand a farming or ranching operation because of neighbors, including many who have moved into the rural community who are able to put a stop to any of those plans.

Let me give you some examples. A few years ago a rural community in Wisconsin wanted to build an ethanol plant. But people who moved out from Madison to retire in the community decided they didn't want an ethanol plant in their town, so they went to court and managed to stop it. Since their retirement income was in place, they didn't need the income the plant would have provided local workers, much less the increased market for corn farmers in the area. NIMBY's won, local people lost.

I've talked before about the community in western Wisconsin where I grew up. Four years ago a wind energy company started to contract with farmers to install turbines and pay them an annual lease fee, providing badly needed income to a suffering farm economy. Well, two retired couples from Chicago who moved there to retire in their idyllic rural home had the money to hire attorneys and file lawsuits to the point where the company will probably abandon its plans and move to another site. NIMBY's win, locals lose.

Then there is the dairy family who wants to build a large dairy farm in northern Illinois and again neighbors, and in this case interestingly enough, not just from towns and cities but farmer neighbors as well, have gone to court to put a stop to those plans. This dairy has met all the state and county rules and regulations and would be operated by a reputable family that has a solid reputation in properly managing dairy farms. In addition, it would provide an excellent local market for farmers in the area who produce hay and livestock feed.

I'm beginning to wonder why states even pass "Right to Farm" laws, because somehow NIMBY's and their attorneys are able to work around them, keeping projects tied up in the courts for years, to the point where farmers give up because they run out of money.

Once in awhile, farmers win and NIMBY's lose. To those who go to court to stop agricultural expansion, take note of a recent case involving a hog producer in Illinois who was sued to put a stop to his expansion. The farmer fought the suit and after several court appearances, the judge ruled in favor of the producer. The judge then ruled that the plaintiffs who brought the suit were liable for all legal fees and are directed to pay those fees estimated at more than $200,000. Finally, some justice!

My thoughts on *Samuelson Sez.*

TELLING AGRICULTURE'S STORY

Agriculture, Every Day in Every Way - March 21, 2009

Every year, National Agriculture Week, the first week of spring, gets an early start in Washington, DC. An important part of that day-long salute to agriculture is the announcement of the winner of the National Agriculture Day Essay Contest. That's an essay contest for students in seventh through twelfth grades. The winning student travels to Washington, DC to present the essay at the kick-off banquet, this year held in the headquarters building of the U.S. Department of Agriculture.

After Secretary of Agriculture Tom Vilsack delivered his National Agriculture Week message, we then met National Agriculture Essay winner Kelly Kohler, a tenth-grade student from Redwood Valley High School in Redwood Falls, Minnesota. The title of her essay: "Agriculture — Every Day in Every Way." I don't have room in this column to share all of it, but let me share some excerpts.

It begins, "A hard-working farmer will be the first person to tell you that agriculture is an essential part in our everyday lives. In small town Minnesota, where a considerable amount of the population has a career in the agricultural industry, this is a widely known and accepted fact. Fields serve as a constant reminder to the inevitable truth that agriculture is a large part of our lives.

On the contrary, in large cities where the view of the fields is hidden by looming skyscrapers, agriculture is practically a foreign concept. Thousands of people from the metropolitan area eat, touch, drink and wear agriculture and for the most part, the urban population is oblivious to the effect agriculture has on their lives.

How are urban dwellers to know that, as they walk down the street, their life is impacted repeatedly by agriculture? The leather briefcase in hand, the burger for lunch and even the shirt on their back are all products of the agricultural industry. The leather and beef came from a cow, the fibers in the shirt from plants that grew on a farmer's property."

She concludes her essay: "Agriculture is everywhere. Even though society seems to have forgotten, it surrounds the countryside, cities and everyone in them. The agricultural industry is one of the most important industries in the world today. A world without agriculture is a world without life."

The winning words of tenth-grader Kelly Kohler, Redwood Falls,

With some Noon Show listeners and Farm Department partner Bill Mason (seated) during a remote broadcast at the International Livestock Exposition, circa 1970.

Minnesota. Congratulations, Kelly and thank you for letting me share your thoughts on *Samuelson Sez.*

Thanks for Your Interesting Thoughts - November 22, 2009

I think I could take four or five weeks to share all of the thoughts you sent to me after I asked you to offer your ideas on how agriculture should tell its story. There are a few more I want to share with you this week and I thank all of you again for taking the time to let me hear your ideas.

Rex in Indiana wrote to say, "I just graduated from high school this summer. I was lucky enough to be born and raised on a farm and around agriculture. My fellow classmates don't have any idea about agriculture and they should be taught. I know that's what agriculture classes that co-exist with FFA are for, but my school doesn't have any of those things. The few of us that are involved in agriculture had to go to another school for those classes. Agriculture should be mandatory for all school kids, if it's at least just one semester in middle or high school, because then the kids would actually gain from the class."

Sandy in Ohio wrote, "Should we relate the evolution of the confinement aspect of hogs and chickens to our urban folk to help them understand why we have moved them into crates or cages? Urban folks would be appalled to see a sow eat her piglets as fast as she gave birth to them. How would they feel if they watched chickens as they peck one of their own to death then move on to the next lowest in the pecking order. As for a spokesman, wouldn't it be nice to find a dynamic young person as a match for the spokesman for the Humane Society of the United States. He is very good."

Finally, I like this one for its humor but I don't think it will work. From Gerald in Illinois: "The Animal Rights people never give up so the rest of the 50 states should, to quote W.C. Fields, 'Take the bull by the tail and face the situation.' It occurs to me that if we use their logic, that livestock abuse happens and therefore we must ban livestock production, then it seems equally logical that the same logic should be applied to human beings. Since children are abused, we must stop producing children."

Many of you offered similar ideas... get involved with urban civic clubs and urban churches, encourage agribusiness corporations to spend advertising dollars that explain farming and ranching (ADM and Monsanto are doing that and plan to increase the dollars), establish county and state agricultural speakers bureaus to get real producers in front of urban consumers, and increase financial support for Ag In The Classroom programs. These are all great ideas; just remember they take time, money and commitment

on the part of all of us, and right now PETA and HSUS seem to have more of all three of those ingredients than we do.

Thanks to all of you for sharing your ideas and again, providing me with your thoughts on *Samuelson Sez.*

Celebrate National Agriculture Week - March 14, 2010

For me there are three very special events in the month of March. First of all, there is the official arrival of spring. That happens March 20. Secondly, National Agriculture Week is March 14th through the 20th this year. Oh, and the third one, my birthday, on the last day of the month.

But this week let's focus on National Agriculture Week, a week set aside for more than two decades now by Presidential proclamation to call attention to the farmers and ranchers of America and the role they play in putting food on our table, clothes on our back, a roof over our head and now, energy in the gas tank.

That is why we celebrate National Agriculture Day on the first day of spring and extend that into National Agriculture Week with many activities across the country, from our nation's capitol to rural communities as well as cities, throughout the week. Many of the week's activities are sponsored by the Agriculture Council of America with assistance from the National Agriculture Marketing Association and financial support from agribusiness firms and commodity organizations.

Why should Americans celebrate Agriculture Day? The Agriculture Council offers several reasons.

* Americans need to understand the value of agriculture in our daily lives.

* Increased knowledge of agriculture and nutrition allows individuals to make informed personal choices about diet and health.

* Informed citizens will be able to participate in establishing the policies that will support a competitive agricultural industry in this country and abroad.

* Employment opportunities for young people exist across the board in agriculture. 4-H, FFA and AFA members can choose from hundreds of careers in farm production, agribusiness management and marketing, agricultural research and engineering, food science, processing and retailing, banking, education, landscape architecture, urban planning, energy and probably some jobs that haven't even been invented yet.

* Beginning in kindergarten and continuing through 12th grade, all students should receive some systematic instruction about agriculture. Agriculture is too important a topic to be taught only to the small percentage of students considering careers in agriculture.

We need to take advantage of the media platform provided by National Agriculture Week, use it to get urban attention and respond to the growing number of attacks on our food producers and food production system. So if you are a farmer or rancher, take a moment to pat yourself on the back, then go out and tell your city friends what you and your family do to feed the world. That's National Agriculture Week.

My thoughts on *Samuelson Sez.*

Let's Take the Country to Town - July 9, 2010

Very often we think the challenges facing farmers and ranchers in the United States are unique to this country. Then we welcome a foreign visitor or we travel to a foreign country and suddenly find that, no, the challenges facing food producers around the world are quite similar.

Recently, Caroline Stocks, a reporter for a major farm magazine in the United Kingdom spent some time in my office. She brought along a copy of the magazine entitled *Farmer's Weekly* and the front-page headlines told me that farmers in England face the same challenges we do: "How the

National Budget Affects Farmers." "Large-Scale Dairies' Public Relations Battle." Among several columns in the magazine, I found one written by a farmer, Guy Smith, who farms about 1250 acres in the United Kingdom, a sizeable farm in that country. His column was entitled "Let's Work Together and Take the Country to Town."

He wrote about a sad event that occurred in the United Kingdom this year. For the first time since 1839, the Royal Agricultural Show near Stoneleigh was not held, cancelled because of low attendance and lack of interest. The show goes back to the days of Charles Dickens and generations of families had attended every year to witness an international display of livestock, horses and agricultural equipment. I attended the Royal Agriculture Show several times in the '80s and '90s and found it to be an absolutely fascinating show attended by thousands of farmers and city folks, as well as members of the Royal Family.

While many people mourn the loss of the show and wish it could be revived, Guy Smith writes "The fact is, now our nation's greatest industry goes without a showcase." So, he suggested a plan that deals with what? Telling the story of English farmers to city people. He said, "Let's take a look at what farmers did in Paris for a couple of days. The Avenue de Champs-Elysees was turned into a breathtaking farmscape. Thousands of French farmers recreated farms in the heart of Paris."

Guy then asks, "Why can't we do something similar in London? It strikes me over the past few years we have become far more proactive in promoting ourselves, but we haven't taken our show to London. With the Royal Agriculture Show gone, it's time we do that," because, he said "We as agricultural producers need to gain more understanding of city people for what we do in this country." Sounds familiar, doesn't it? Since agriculture and its challenges are truly global, perhaps we can find some global solutions and different ideas to respond to the attacks on agriculture. Let's work on it.

My thoughts on *Samuelson Sez.*

Ag Communication Must Improve - January 29, 2010

I have lost count of the number of times over the years that I've heard people at farm meetings say, "We must do a better job of telling the story of agriculture to consumers." As a matter of fact, I have lost count of the number of times I have said that on Samuelson Sez. But I say it again and I say it this time from a little different approach.

First of all, the need for communicators who understand agriculture is more urgent than it's ever been. The number of attacks on animal agriculture and on the way we produce food continues to increase every year. We need knowledgeable people with agricultural backgrounds, to work not only in agricultural media, but perhaps more importantly, in urban non-agricultural media. We also need anybody and everybody involved in food production today to tell the science-based story at coffee shops, supermarket checkout lanes, PTA meetings and Sunday morning church coffee hours. And that means each of you!

But let's focus for a moment on the professional communicators. Where do we find the new crop of agricultural broadcasters, writers and speakers to counter the attacks on agriculture? At the recent trade show in Orlando, Florida, Ag Connect Expo, there were thirty information sessions dealing with various topics. I was involved in one of them, a three-person panel looking at agricultural communications in the future. Joining me on the panel were Mike Yost, a South Dakota farmer who has traveled the world as President of the American Soybean Association and then as head of the Foreign Agricultural Service of USDA, and Dr. Jim Evans, retired Agricultural Communications Professor at the University of Illinois.

It was special having Jim on the panel because he has touched the lives of so many of us in agricultural communications during his more than three decades of teaching. Because of his dedication to the profession, many of his friends across the nation and the University of Illinois are conducting a drive to raise the money to establish an Endowed Chair for Agricultural Communications in the name of Jim Evans. Currently there is no Endowed Chair for Agricultural Communications at any university in the country.

We need to strengthen the educational process for agricultural communications to find and recruit students who have the talent and desire to do the work that needs to be done and then to provide them with the quality education that will help them tell the true story of agriculture. It is our hope that establishing an Endowed Chair at one university will lead other land-grant colleges to do the same to lift the importance of the program and prepare the next crop of agricultural communicators. It is critically important to all of us in the industry and we need your help and support.

My thoughts on *Samuelson Sez.*

Are We Making Any Progress? - September 5, 2011

Are we making any progress? It's a question I frequently ask. Are we making any progress in telling the story of American agriculture to the 98% of the people in this country not involved in farming or ranching.

As I moved through the 75-acre exhibit field at the 2011 Farm

Progress Show in Decatur a few days ago I again marveled at the technology and equipment being developed by agricultural engineers, computer specialists and plant geneticists to improve the ability of America's farmers and ranchers to feed the world. It is impressive and while I understand that exhibitors want to talk only to potential customers, it's unfortunate that city people don't have the opportunity to get a better understanding of the technology and equipment producers use to put food on their table, clothes on their back, a roof over their head and now energy in their tank.

Since 1960 I have had the opportunity to talk to hundreds of thousands of city people in the Chicago area on WGN Radio, using the 50,000 watts of power to educate them about agriculture. I think I have had some success, but the reason I continue to ask the question, "Are we making any progress?" is because I'm still getting questions that we really shouldn't have to answer anymore. Questions like:

- Why do farmers put poison on their fields and over-use antibiotics in their livestock?
- Why do we export so much of our food to foreign countries because if we didn't, food in this country would be much cheaper?
- Why are all laying hens kept in cages and livestock kept in stalls or pens instead of being free to enjoy the outdoors on green grass under blue skies?
- Why should my taxpayer dollars go to pay farmers not to produce?
- Why can't farmers stop polluting the air and the water?

These are all questions that shouldn't have to be asked by city people anymore. We try. Over the years I have watched efforts by the National Farm-City Council, the Agricultural Council of America, which is the current alliance to tell agriculture's story, and the Agriculture in the Classroom program. All well-intentioned, but underfunded, in their efforts to answer the questions and build a better understanding of, and appreciation for, America's farmers and ranchers.

The message here is that consumer education is a never-ending job because every day new people arrive on the planet and are another generation removed from the farm. That's why each of us involved in any phase of production agriculture must take every opportunity to help people understand that farmers and ranchers need the technology and the equipment so they can continue to improve production and feed this growing world population that's expected to hit nine billion people by 2050. The job is huge but we must do it to ensure a much more positive answer to my question.

My thoughts on *Samuelson Sez.*

MAD COWS

Let's Stop Overreacting to Issues - October 10, 2008

I considered it to be a significant event at the 42nd Annual Dairy Expo in Madison, Wisconsin earlier this month; the return of Canadian cattle to the show ring and the judging arena. The Canadians were happy to be back as well, as they certainly should be, in light of the fact that the highest priced animal in the World Classic 2008 Holstein sale was a Canadian Holstein that brought $97,000 and the Supreme Champion of World Dairy Expo over all breeds was a Holstein from Canada. Yes, they did very well.

But, let's think about recent history here for a moment because this was the first time Canadian cattle were allowed to cross our border and compete at World Dairy Expo since the 2002 Expo. It was May 20th of 2003 that North America discovered its first case of Mad Cow Disease in Alberta, Canada. Later that year, December 23rd, we found our first case in Washington state. The result: borders closed around the world to the movement of livestock, both dairy and beef.

Talking to the Canadian dairymen at the show, they said it had a tremendous negative impact on their business. Mark Comtrois from Quebec, Canada, is a major player in genetics and pedigree sales around the world and he said his business dropped 85% that first year. He had to totally realign his dairy business, relying less on sales of genetics, and buying additional Canadian milk quotas to maintain income and stay in business. He told me many of his fellow dairy producers in Canada did not survive.

Which brings me to a point that I've made several times: The Mad Cow Disease story is the biggest non-news story I have covered in all my years of reporting agricultural news. Compare the dire predictions and the end results. Some experts predicted at the time that 50,000 people world-wide could die of the human variant, Creutzfeldt-Jakob Disease which is believed to be related to Mad Cow Disease. At last count, less than 200 people had died of Classic CJD which is not related to MCD. That doesn't look like much of a threat to me. Livestock authorities have taken the necessary steps to control the problem, so why should borders still be closed?

One more point... same subject, different species. What about the bird flu scare? Experts said 2,000,000 people could die, countries including the U.S. have set aside billions of dollars to fight the impending pandemic. But the world-wide death toll over the past two years? Again less than 200 people.

The stories have received far more attention in media than they deserve and unfortunately, have had a very negative impact on livestock and poultry producers around the world. We need to be better judges of the significance of stories of this type.

My thoughts on *Samuelson Sez.*

DO NOT CALL IT "SWINE FLU!"

Keep Things in Perspective - April 30, 2009

Once again, America's food producers have received another harsh reminder of how vulnerable they are to events that are totally out of their control; events that determine the future of their industry and certainly their financial well being. I'm talking about what I consider the over-blown media coverage of the H1N1 flu virus. That is the correct name, but around the world it quickly became known as "swine flu," and what a negative impact that has had on the pork industry!

Agriculture has a decade of lessons on events being blown out of perspective by ignorant people in media. At the beginning of the decade cattle producers were slammed by media coverage of "mad cow" disease; the middle of the decade it was the turn of the poultry producer, and the exaggerated coverage of "avian bird flu;" and now, pork producers, it's your turn with "H1N1" and the extensive coverage it has received around the world.

Despite continued assurances from medical doctors and agricultural media that you do not get "swine flu" from eating pork, consumption has declined. Futures prices moved limit down at the Mercantile Exchange; cash hog prices dropped $7-$8 per hundredweight the first week, hurting pork producers at a time when they were already suffering from prices below the cost of production. Beyond that, major importing countries of U.S. pork, China and Russia shut the door on pork imports from North America, knowing full well that H1N1 is not spread by consuming pork and their action violates world trade rules.

I tip my hat to the European Union for taking quick positive action. Expressing concern for the European pork producers, the EU Health Commissioner said on the second day "Not to have a negative effect on our industry, we decided to call it 'novel' flu, not swine flu from now on."

Why do I think this is another non-story getting far more attention than it deserves? Every year 35,000 people in America die of influenza and we don't interrupt programs with "breaking news" to announce those deaths.

Finally, some history. Let's go back to 2005 and the bird flu story. In my files I found an Associated Press story quoting a United Nations health official who said, "Bird flu could kill 150 million people on the planet" and countries around the world spent billions of dollars to prepare for the pandemic. I checked World Health Organization records and found that, as of April 23rd this year, on the planet, 257 people have died of bird flu, hardly a pandemic. Let's work to keep these stories in perspective, so we don't kill another agricultural industry.

My thoughts on *Samuelson Sez*.

Let's Help Media Get it Right - June 6, 2009

I'm really concerned that I'm going to have to start taking blood pressure medicine if my fellow journalists don't start practicing accuracy in dealing with the H1N1 flu virus. Every time I see a television news story about "swine" flu illustrated with video of hogs on a Midwestern farm, my blood pressure goes right through the roof. And you know, I'm not alone; as a matter of fact, by talking to pork producers, I know that it does the same to you and it's so frustrating that we don't seem to be able to do anything about it. Newspaper writers, radio and TV reporters continue to refer to it as swine flu even though there is no connection between eating pork and getting H1N1. The World Health Organization, the U.S. Department of Agriculture, the Center for Disease Control have all issued statements saying it is not "swine" flu, it is H1N1 Flu Virus.

Someone else who feels as strongly is Secretary of Agriculture Tom Vilsack. At a farmer forum in Illinois recently, the Secretary was asked about it by a young FFA member who said she and her family were pork producers and she wondered what they could do to get media to deal accurately with H1N1. The Secretary's response was so strong that we carried it on our TV show, *This Week in AgriBusiness*, and I want to share it with you.

Secretary Vilsack said, "It is so irritating when you read the paper, when you watch on television, and they continue to refer to it as 'swine' flu. Because what they don't realize; what that good-looking, handsome, well-coiffured talking head on the television doesn't understand is there is a hard working family whose livelihood depends on him or her getting it right. They don't understand that we saw as much as a $750 million potential hit to the pork industry because other countries made the decision, without science, without basis in science, of saying, 'We don't want your pork products.'"

The Secretary was as frustrated by the inaccuracy of the media as

Mike Perrine and Don Tuttle worked with Max and me on the Tribune Radio Network.

the rest of us and I appreciate his strong words of support. As a journalist, I am always concerned about being accurate when I report a story and I don't understand why my fellow journalists ignore accuracy on this story.

Now, I do have a suggestion for that Illinois FFA member and the rest of us who see the financial impact of the "swine flu" reference on pork producers. Every time you see or hear swine flu in a story, fire off an e-mail or note to the editor or news director and suggest, in the interest of accurate reporting, they call it what it is, H1N1 Flu Virus, not swine flu. It will just take a moment, but I think it is time well-spent.

My thoughts on *Samuelson Sez.*

ANIMAL TRACKING

Traceability: It Is Important - October 1, 2006

The recent E.coli in spinach situation raises several concerns. First of all, of course, the concern over human health and the tragic loss of life.

Secondly, the concern over the economic loss for people who have put their life's work into producing top quality, safe produce; then the issue of how vulnerable we are to bio-terrorism and finally the need for traceability, the ability to trace our food all the way back to its source.

The bio-terrorism issue is certainly there. Security experts have said, "Terrorists don't need to explode a bomb in the United States; all they need do is bring a vial of foot-and-mouth disease or some other contagious virus to contaminate our food supply and we'll be brought to our knees." As we look at the issues, and particularly traceability, we must focus on the national animal ID program. That always sparks an emotional debate, particularly with those people who don't like the idea of "big brother" or having to do more bureaucratic paper work.

Yet I am convinced the food industry and consumers are going to demand traceability, whether it be produce, meat, milk or eggs. And in the competitive global market, our foreign customers are going to demand it. In the spinach situation, it would have been extremely beneficial to be able to immediately go to the database and find the source of the contaminated produce so we could stop sales from the infected source, but allow sales to continue from areas that proved to be free of E.coli. That would have been good for consumers; it certainly would have been good for producers. That is why, when we look at animal identification, we must look beyond the producer skepticism of government officials and bureaucrats and realize the long-term benefits for all of us. Like it or not, we live in a different world today, and traceability becomes important; to our health as well as our war on terrorism. Traceability is not something new; I know producers who are under contract to provide ingredients for companies who produce baby food. They are required to keep detailed records of everything put on every acre of carrots or beans so in the event of a recall, the company can immediately go right to the acre and find the source of the problem.

In animal agriculture, the day is rapidly approaching when packing companies and their customers like McDonald's and Burger King will not buy meat animals from producers who cannot guarantee traceability. So, let's get over it. Let's find a way to get it done, ideally, on a voluntary basis with the cost to be shared from the producer all the way to the consumer. Let's not ignore the lesson to be learned from the spinach situation.

My thoughts on *Samuelson Sez.*

America Gets COOL - July 31, 2008

Just in case you have forgotten, September 30th is the date that

America gets COOL. That's right! COOL... Country of Origin Labeling mandated in both the 2002 and 2008 Farm Bills. It is finally here, and a few days ago, the Department of Agriculture issued an interim final rule for the mandatory Country of Origin Labeling program.

Some of the highlights of the rule; it covers muscle cuts and ground beef, including veal, lamb, goat, chicken and pork meat; perishable agricultural commodities, fresh and frozen fruits and vegetables; macadamia nuts, pecans, ginseng and peanuts, as required in the original legislation. Commodities covered under mandatory COOL must be labeled at retail, to indicate their country of origin. However, they are excluded from the program if they are an ingredient in a processed food item.

Food service establishments such as restaurants, lunchrooms, cafeterias, bars, lounges and similar enterprises are exempt from the mandatory country of origin labeling requirements. Generally, reaction has been positive; as a matter of fact, it brought together some strange bedfellows in support. The National Cattlemen's Beef Association said in its statement, "We are pleased to have an interim final rule in place. It incorporates provisions that make mandatory labeling more feasible for producers." Then R-CALF, which seldom agrees with NCBA on anything said it was "pleased to see the new interim final rule; for the most part a simplified and workable rule for independent U.S. cattle producers."

That's all well and good, but for me it begs the question... How do we trace these foods to their very beginning without a national and workable traceability program? I think it comes back to something I've said before (and a lot of people disagree with me); we need a national animal ID program. Now I know COOL and the NAIS program are two separate pieces of legislation, supposedly not related, but how can we have one without the other? How could we trace a food problem to its origin if we don't have better traceability than we have now?

I come back to the recent tomato salmonella scare. An entire segment of agriculture, tomato production, was brought to its knees and growers lost millions of dollars because we had no quick way to find the source. Adding insult to injury for tomato producers was the final determination that the source was jalapeno chiles from Mexico.

My point is until we can quickly find the source of contamination, the entire industry and all of its producers will pay the price. Until we have a real traceability program in place, I doubt that we can really be COOL!

My thoughts on *Samuelson Sez.*

At a 1999 speaking engagement with Tom and Pat Strachota. Tom is CEO of Dairyland Seed in West Bend, Wisconsin and Pat is a member of the Wisconsin State Assembly.

Unintended Consequences Can Harm - July 25, 2009

"Unintended consequences." That's a term that was used a few days ago by Gary Voogt, the President of the National Cattlemen's Beef Association, when I interviewed him at the Summer Cattle Conference in Denver, Colorado. We were talking about COOL, Country of Origin Labeling. Gary is not a big fan of COOL and said when it comes to reading labels, he's not convinced that many consumers take the time to do it. If they do read the label on a package of beef in the supermarket meat counter, he said they will find that 90% of it is labeled "U.S. Beef."

He said, on the other hand, it has created some trade problems for us. Pointing out that Canada and Mexico, who are large buyers of American beef, now look at country-of-origin labeling as an artificial trade barrier and are filing complaints with the World Trade Organization. That, he said, is an unintended consequence of COOL.

That got me to thinking about other examples of unintended consequences and one jumped out immediately. The people who succeeded in passing legislation to ban the slaughter of horses in this country thought it would improve conditions for aging horses and keep them from "suffering an inhumane death" in a slaughter plant. Yes, death in any form is unpleasant, but look at how these horses now die; turned loose by owners who can no longer afford to keep or feed them or dispose of their carcasses, these horses are now subjected to disease and predators and ultimately die of starvation in the countryside. With the tough economy today, more owners are turning their horses loose, to the point that horse shelters are filled to capacity. So, backers of the bill, tell me which is better for the aging horse: humane slaughter in a plant regulated by USDA inspectors or starvation in the country. I know which answer I support and to me, this is one more example of an unintended consequence. I might add that I haven't heard any of the Hollywood stars who supported the bill volunteer to spend their millions to take care of abandoned horses.

Finally, I asked you to share with me your feelings "for" or "against" on the national animal identification issue; I'm hearing from a lot of you and I will be sharing many of those quotes in an upcoming column.

I do have a request of a few of you; watch your language. There is more than one side to this issue and you don't have to use strong language to disagree with me. I respect your position and I ask that you respect mine. I am not "stupid" or "in the pocket of corporations or government," or the unprintable words that some of you used. I think we need to mind our manners and whether it's me or other media people, to have your opinions heard and seriously considered, it helps if those opinions are voiced in a civilized manner. Otherwise, you may find yet another "unintended consequence."

My thoughts on *Samuelson Sez.*

Traceability is Coming - October 10, 2010

How well I remember a couple of years ago, the letters and e-mails I received from livestock producers across the country who were angered when I called for a national animal identification program. Those were angry and downright nasty e-mails. But, it is time to say once again, that traceability in the food industry is coming, whether you like it or not.

Processors, retailers and consumers are going to demand traceability because of the increased urban media coverage of food recalls; the most recent, of course, being the salmonella egg recall that sickened 1,600 people and led to the recall of 550,000,000 Iowa eggs. Until the source was found, the entire egg industry was condemned by consumers, many of whom stopped eating eggs because of their concern over safety.

The beginning of the demand for traceability goes back to 2006. Remember the E. coli contamination of a California 50-acre spinach field that led to five deaths and 200 illnesses in 26 states? That spinach was canned under the Dole label and that is when Dole contracted InSync Software to set up a computer electronic tracking system similar to FedEx that now tracks every crate of produce from the farm field to the supermarket.

IBM is now working with a group of leading California growers associations at their request; Intelleflex is working with food and vegetable producers in Hawaii; YottaMark is working with 2,000 yam and berry producers and Infratab is working with California grape growers setting up computer programs that can trace every product from the field all the way to the dinner table. These tracing programs ultimately will benefit agricultural producers of every commodity, including livestock producers, because if the source of contamination can be located and isolated immediately, then you don't condemn the entire industry.

Traceability is already here in some agricultural commodities and will eventually cover all food producers. All it will take is for a Walmart or Costco to announce they will no longer buy meat or other products that cannot be traced back to the farm and that will push everybody to scramble to meet their demands. My hope is that processors will offer financial incentives to encourage producers to adopt tracing programs without placing the entire burden of cost on the producer.

Other countries are far ahead of us in this technology and we should learn from their experience in establishing the programs with their producers and satisfying the demands of processors and consumers. Let's get ahead of the curve because, like it or not, the demand for traceability in the food industry is here.

My thoughts on *Samuelson Sez.*

PETA, HSUS AND OTHER NONSENSE

It's Back to the Courtroom - September 29, 2009

If you are a producer of poultry or livestock, it seems the challenges these days are never-ending. People in organizations all over the planet want to put you out of business. From the *TIME* magazine article a few weeks ago, filled with misinformation that accused livestock and poultry producers of polluting the environment and the world, to the latest attack on the industry brought to my attention by Gary Baise, former Illinois farmer and now, in my opinion, the country's leading agricultural environmental attorney based in Washington, DC.

Gary informed me of a new attack that has been launched by the

Humane Society of the United States, the Association of Irritated Residents, Friends of the Earth and other environmental groups. On September 21st they filed a sixty-eight page petition with the Environmental Protection Agency, which, if implemented, could achieve the environmental organizations goal of shutting down concentrated animal feeding operations, CAFO's, in the United States.

Gary pointed out that in the petition, environmental groups want to control, among other things, the diet you feed to your animals, the pH of manure, as well as the time and temperature that the animal waste is in storage. They also want you to switch from farm animal production systems that are reliant on grains to systems that use pasture-raised, organic or full-cycle farming systems.

The groups are using a legal device to request and possibly force EPA to regulate CAFO emissions under section 111 of the Clean Air Act. The reason they are seeking regulations in this section is they believe CAFO operations are producing large amounts of air pollutants that endanger the health of humans, animals and ecosystems that are far removed from these CAFO operations. Another concern about using this section is that it not only impacts new sources of pollution, it can be applied to existing sources as well. Attorney Baise concludes, "This petition by the Humane Society is a much more dangerous legal tactic because the Clean Air Act does not require an absolute certainty of proof of actual harm when making an endangerment finding."

So, it's back to the courtroom and I truly wonder if the leaders of these groups are intelligent enough to realize that the more money and time food producers spend in the court room, the less time and money they have to produce food, which means more people in the world go to bed hungry every night. I didn't even know there was an Association of Irritated Residents, but I'm ready to join so I can express my extreme irritation with them and the other groups that get in the way of feeding the world!

My thoughts on *Samuelson Sez.*

No Farms, No Food...Think About It - October 9, 2009

The growing number of attacks on America's livestock producers that I've discussed on *Samuelson Sez* the past couple of months has generated a great deal of response from you, expressing your feelings and concerns about the future of animal agriculture. I'm grateful for your response and I want to share just a couple of excerpts on this week's *Samuelson Sez.*

In an earlier commentary, I discussed "State Issue Number 2" that

Courtesy: Don Peasley Photography

Angelo Lazzara gets some close-up video of me hooking a cow up to a milking machine at the Harvard Milk Days in 1998.

will be on the ballot in the state of Ohio on November 3rd. A YES vote will mean that livestock producers will determine how they raise and care for livestock; a NO vote would mean PETA and the Humane Society of the United States would decide how producers care for livestock.

An Ohio farm couple sent me a two-page letter to express their concerns. I quote a couple of lines from their letter: "The consequences of doing nothing and letting these people have their way will be devastating. Agriculture as we know it today will cease to exist. The uninformed, no-knowledge people have no idea what it takes to operate a farm. I doubt if any of them have ever lived on a farm."

The basic purpose of the letter is to call for Ohio agricultural people to be as informed and vocal as possible, urging all citizens to vote YES. I like their closing line: "Remember, the squeaky wheel gets the grease; do nothing and you will get the shaft."

Then, a viewer of our TV show in Maryland wanted me to know that this month, the Baltimore Public School System became the first in the U.S. to pledge to serve no meat in school lunches on Mondays. According to the organization behind the campaign called "Meatless Monday," that means 80,000 students have no meat options in the school lunch menu on Mondays. The "Meatless Monday" campaign is committed to cutting out meat one day a week for their health and the health of the planet. I can never understand how the health of the planet will be improved if we don't eat meat one day a week. It is simply another attack on livestock producers and an extension of the campaign to turn all of us into vegetarians.

Finally, I thank the person who sent me a copy of Norman Borlaug's obituary, printed in the *Economist*, that said Mr. Borlaug called these people who are against any technology to increase food production "naysayers and elitists who have never known hunger, but think for the health of the planet, the poor should go without good food."

My closing comment comes from a farmer who said: "We should put a label on every package of food in every store in America. The label should simply say 'NO FARMS, NO FOOD!'" I agree!

My thoughts (and yours) on *Samuelson Sez.*

A Listener Makes a Point - August 13, 2010

I do enjoy reading your responses to some of my topics on *Samuelson Sez.* I will tell you however, if your name is not on the e-mail I don't even read it, but let me share with you an e-mail this week from a reader who used his name and is a Doctor in Environmental Sciences.

"Dear Mr. Samuelson,

The reason I am writing you today is that you regularly lump 'environmentalists,' 'animal rights advocates,' and 'environmental scientists' into the same bag. This is highly inappropriate. There is a huge gulf between those who work on environmental issues, those who tie themselves to trees and those who wish to give animals the right to vote.

In fact, most scientists will agree with agriculture professionals that groups like PETA and Humane Society of the United States are problematic groups. Also most environmental scientists have very little in common with radical environmentalists and animal rights advocates.

We do not have it out for agriculture and definitely desire to find ways for agriculture industries to adopt more environmentally friendly practices while improving their profit margin. Agriculture uses natural resources and every farmer I have ever spoken with possessed extremely positive environmental ethics. Posing those environmental scientists who seek to help farmers as members of radical groups is simply inappropriate."

Yes, perhaps I have painted some people and groups together with too broad a brush and for that, I apologize. I don't recall referring to environmental scientists but I certainly agree with this Doctor of Environmental

Sciences that there is a difference between what I call the radical elements in environmentalism and animal rights and those who, as this Doctor says, understands agriculture and the need to make a profit and works to enhance environmental practices.

Doctor, I thank you for calling it to my attention and in the future will make it a point to refer to the PETAs and HSUSes of the world as radical and will not include environmental scientists working with crop and livestock producers to improve the environment. I would also urge all of us who become involved in this debate with the radicals, to always use science and not emotion as the basis for our arguments.

My thoughts on *Samuelson Sez.*

Fight the Foe, Not Each Other - October 1, 2010

We have an ongoing situation in the agricultural community in this country that really bothers me. I am talking about the fighting inside the agricultural community; producers fighting with each other over the quality, the nutrition and the safety of the food produced on their farm or ranch.

We cannot afford the infighting when attacks from outside the agricultural community continue to grow. We must respond to the radical environmental groups, the Humane Society of the United States, PETA and other organizations that would totally alter the way we produce food in this country, to the point where perhaps in the future, we couldn't produce enough and would have to import food from other countries.

Yet, e-mails and letters, headlines and stories continue to come across my desk. Organic farmers claiming their food is better than what they call conventional food; locally grown food vs. food that travels a long distance to get to our homes; family farms vs. factory farms. All of this infighting is costly; it also raises questions and lowers confidence about the safety and quality of our food supply in the minds of consumers.

Every time I mention organic farming, I hear from organic farmers saying, "Why are you against organic farming?" I am not! As a matter of fact I tip my hat to those organic farmers who have found the niche market and consumers who will pay a great deal of money for organic food. But I get upset when organic producers make claims that their food is better, safer and more nutritious; because there are no scientific studies to support that claim and, in effect, it attacks non-organic producers.

Then there is the "buy locally" campaign, which is fine, but don't intimate that's better for the environment than food that travels thousands of miles. In the winter time if you live in the northern part of the country you cannot buy locally grown fruits and vegetables, they must travel long distances from California or Florida or we go without.

Then there is the term corporate farm that implies the operator is some faceless person only interested in money at the expense of the environment and animal care; even worse is the term factory farm. To me, a factory is an assembly line with workers building a product as it moves down the line; that is not what I see when I visit farms, large and small. I've learned over the years that every large farm started as a small farm and more than 90% are family corporate farms.

Big farm, small farm, organic, non-organic, local, long distance, we are all in the same business, feeding hungry people in a growing world, and there is room for all of us in this business. We must stop the infighting in our agricultural family and instead, use our resources, time, energy and money, to respond to those who attack us from the outside.

My thoughts on *Samuelson Sez.*

You've Been Doing It Wrong, All These Years? - November 19, 2010

So you are a producer of poultry and livestock that ultimately goes to market and ends up on dinner tables across the country. You are doing a good job; you are concerned about the nutrition and the health and the well-being, and the humane treatment of your birds and your animals and you follow the humane programs put forth by animal scientists at land- grant colleges and universities. You work hard at it.

Well my friends, I'm sorry, you are not working hard enough. A story a few days ago in the *Chicago Tribune* by reporter Monica Eng started this way:

"Whole Foods Market harbors the same hopes for its chickens that many parents do for their kids; that they will get plenty of fresh air, live at home until they reach maturity and avoid gaining weight so fast that they can't walk. These are a few of the animal welfare practices the retailer hopes to encourage with a humane meat rating system being piloted in the South and scheduled for national expansion early next year."

Whole Foods Market will add another label to that package of meat or poultry and it is the "animal welfare rating" label. Now, listen to what you, as a producer, must do to get the highest animal welfare rating to sell at Whole Foods Market.

- Broiler chickens: They must be bred, hatched and raised on farms under the same proprietorships. When moved, they must be carried upright, one at a time. Maximum group size, 500 birds; and they must be able to perch.

- If you are going to produce pork that will get you the highest animal welfare rating label at Whole Foods Market. Slaughter is required at the same or a local farm; pigs must remain with their litter mates for their entire lives; they must have unrestricted access to mud wallows and there must be vegetation on at least 50% of pasture.

- If you are raising beef, to get the highest animal welfare rating: No electric prods, antibiotics or growth hormones; slaughter required at the same or local farm; cattle brands and clipping ears for identification prohibited; no castration allowed; and calves must be naturally weaned from their mothers.

Here we go again with more rules and regulations, put in place by people who more than likely have never raised poultry or livestock. Nowhere in the story did anyone from Whole Foods Market say anything about paying producers more if you do all of these things, but they did say they are going to work to attract other national chains and restaurants to join them in their "animal welfare rating" program. Where will it end?

My thoughts on *Samuelson Sez*

Humane Society of the United States is NOT Involved in Local Humane Societies. Learn the Difference! - December 10, 2010

Some of you have criticized me for my criticism of the Humane Society of the United States; I will quickly tell you this is a point on which you and I will never agree. I am critical of the HSUS because I think it asks for your financial contributions under false pretense and I thoroughly disagree with their stated goal of shutting down animal agriculture in the United States.

I am not alone in that criticism. A listener recently sent me an article from the *Daily Nebraskan*, the student voice of the campus of the University of Nebraska in Lincoln. The article was written by Jake Geis, who is a second year veterinary student. His story was headlined "Humane Society of the U.S. Has Backward Priorities" and let me quote from the article: "Local humane societies are wonderful things, one of the more underappreciated organizations in America. They provide a second chance to many pets that otherwise would have been put down or left to starve. Unfortunately, the Humane Society of the United States is not one of these. It is NOT the national office of our local humane societies, but rather an activist organization whose main goal is to end animal agriculture, hunting and research."

He points out that the $450,000 in grants the HSUS gave to animal shelters across the country in 2008 amounted to just one-half of 1 percent of the total budget of $99,700,000. He said, "HSUS gave absolutely zero dollars to Nebraska animal shelters between 2006 and 2008."

In his article, he said the two main recipients of your contributions are campaigns to build the "National Rifle Association of animal rights groups" to shut down animal agriculture, and money committed to salaries and benefits for executives and staff of HSUS. "According to its 2009 IRS Form 990, the not-for profit group tax form, HSUS had a revenue of $101.6 million, and spent $35.8 million on salaries and benefits, $2.6 million on pensions and $22.3 million on advertising. Grants to actual humane societies were less than two percent." I'm grateful to Jake Geis for his work in sharing the facts on HSUS .

The bottom line is this, if you want your dollars to go to pet and animal care, then contribute directly to your local animal shelter and know that your money will not be spent for other purposes.

My thoughts on *Samuelson Sez*

Sharing Thoughts from Listeners - December 29, 2010

My year-end editions of *Samuelson Sez* always originates from Scottsdale, Arizona where I annually spend my Christmas and New Year's vacation. And since time in the desert climate tends to make me lazier than normal, I'm going to let you write this column by sharing some of your reactions to *Samuelson Sez.*

The "unwanted horse problem" generated many e-mails. A reader in Colorado wrote "People that are against the slaughter bill have NO IDEA where all the unwanted horses go and how the bill has affected the horse market. Also, if people have the idea that ending USDA-controlled slaughter in the U.S. has taken care of their idea of being humane, they are definitely WRONG! They should visit slaughter houses in Mexico and see the "humane" treatment there and the meat inspection procedures. Sanctuaries for these unwanted and crippled horses is NOT an answer."

My comments on the humane processing rules demanded by Whole Foods Markets of livestock and poultry producers brought this response from a New York reader. "Your comments regarding the business practices of Whole Foods being too restrictive leads me to shout back to you, 'SO WHAT?' If people want to spend their money on animal products produced a certain way, what business is it of yours? If Whole Foods did not sell products that the surrounding community wanted, most likely the store would go out of business. I have realized that you take issue with produce or livestock that are grown or produced organically. Listen, diversity is not a bad thing. Organic farmers have the right to exist, just as the mega-conventional farmers do. If you do not like organically grown products, then do not buy them. That is your choice."

Finally, my comments on a country Lutheran church leaving the Evangelical Lutheran Church of America because of a perceived anti-biotechnology statement by the ELCA brought this response from ELCA headquarters. "The draft does not advise farmers about farm practices, nor does it advise anyone else in biotechnical fields about how to use these techniques. It is a document for discussion of these practices from a theological perspective, and it offers some cautions about the ethical uses of such techniques. It is regrettable that a congregation chose to leave the church over this issue. Apparently, the members misunderstood the purpose of the document." I have been invited to interview Presiding ELCA Bishop Mark Hanson on the draft proposal and will do so early in 2011 and will share it with you.

Thank you for helping me write this edition and these are your thoughts on *Samuelson Sez.*

UNWANTED HORSES

Horses Need Our Help — Humanely - November 12, 2010

I continue to get e-mails from people very concerned about what is happening to abandoned horses. The number of abandoned horses has grown sharply since Hollywood movie stars were able to convince state legislatures as well as Congress to ban the slaughter of horses and the export of horse meat from the United States.

With the downturn in the economy, many horse owners are finding it impossible to properly care for their horses. Rather than have officials find those mistreated animals on their farms and ranches and be criminally charged, they are hauling them away, turning them loose in state parks and public forest land where the animals ultimately die of exposure, starvation, disease or are killed by predatory animals. I would ask the Hollywood stars and animal rights groups if this is really a better way for a horse to end its life instead of going to a USDA-approved slaughter plant, being processed into meat and then exported to foreign countries where horse meat is part of the human diet? I don't think so.

Part of the argument that bothers people is the idea of consuming horse meat. But horse meat is part of the diet in many countries and cultures of the world and there have been times when it has been served in

For 31 years at WGN Radio and TV and many more years (still counting) on "This Week in AgriBusiness," Max Armstrong and I have done our best to tell agriculture's story accurately and encourage others to do so, too.

this country. I remember eating horse meat on our farm when other meats were rationed because they were needed to feed our troops in World War II. And that occurred in cities as well as on farms.

Checking my 20th Century history book, I found this story. "September 27, 1946 - New Yorkers are eating horse flesh in increasing amounts, it was learned yesterday as supplies of standard meats stayed at a record low, black marketing spread and poultry prices soared to $1.00 a pound. Ceiling prices on choice cuts of horse meat are 17 and 21 cents a pound. Former Mayor LaGuardia has called the eating of horses a sign of degeneration, while Health Commissioner Weinstein says horse meat is 'as nutritious and as good as any other meat.'"

I certainly am not advocating or suggesting that you eat horse meat, but I am strongly suggesting that humane slaughter is a much better way to end the life of a horse than the way we are doing it now. It gives owners who can no longer care for their animals in a humane way a much better alternative than turning them loose; and it provides food for hungry people in other parts of the world.

Perhaps with the change of leadership on the House side of Capitol Hill, legislation could be re-introduced to deal with this problem in a realistic way. I think it is worth a discussion with your legislators because I certainly don't see the Hollywood opponents and their followers coming up with the billions of dollars it would take to provide managed "dying farms" for these unwanted horses.

My thoughts on *Samuelson Sez.*

The U.S. Horse Industry Needs Your Support - September 30, 2011

To those of you who sent me the very nasty, angry e-mails about a year ago when, in this column, I supported revival of the "Horse Humane Slaughter" program, I have news for you. Things haven't gotten any better for those unwanted, abandoned horses; they have gotten much worse.

The program was heavily lobbied by several celebrities who testified before Congress and state legislatures to stop the slaughter of horses, the processing of the meat and the exporting of the meat to those countries where horse meat is a part of the culture and diet. They succeeded and the processing plants were closed.

That has not improved conditions for abandoned, unwanted horses in this country. Abandoned for whatever reason; the owner can no longer afford to keep them or feed them and cannot afford to dispose of them properly and so they are turned loose to die of starvation, disease or predatory animals that devour the unwanted horses.

That is why, once again, I'm supporting a new campaign to overturn that ban. It is an effort being conducted by the United Horsemen, a nonprofit that is dedicated to the well-being of horses and horse people. According to its newsletter, "We work for the restoration of humane and regulated horse processing in the U.S. to ensure the best possible fate and valuable use of excess, unwanted and unusable horses. We also work for the responsible management and control of wild and feral horses on federal, state, tribal and private lands. We seek to preserve the private property rights of all horse owners, our beloved horseback culture, and to ensure the long-term viability of the horse industry to not just survive but to thrive economically, socially

WGN Radio

It took a while for Wally Phillips to warm up to the importance of farm reporting on WGN Radio, but when Illinois Farm Bureau members donated about $30,000 from their 1978 corn crop to the WGN Neediest Kids Fund, Wally joined me in a harvest-ready field.

1998 with Wally Phillips, Bill Bell and Lee Phillip Bell, a former Chicago television broadcaster. The Bells wrote and produced CBS daytime dramas "The Young and the Restless" and "The Bold and the Beautiful."

Courtesy: Roy Leonard

WGN legend Roy Leonard and his late wife Sheila in the mid-'90s.

Longtime WGN Radio News Director Tom Petersen and wife Maria at my 50th anniversary party, 2010.

Uncle Bobby's wife,Christine Collins and former WGN "Life After Dark" hosts Johnnie Putman and Steve King, 2010.

Former "Country Fair" and "Noon Show" singer Elaine Rogers sang for me at my 50th anniversary party in 2010.

Max Armstrong and I on the set of our TV show, "U.S. Farm Report."
In 2005, we started another TV show, "This Week In AgriBusiness."

With former WGN host Steve Cochran and WGN news anchor Andrea Darlas.

A promotional piece, circa 1998.

WGN Radio

Uncle Bobby and I "interviewing" each other.

WGN Radio

One of countless remote broadcasts with lawn and garden expert Jim Fizzell, a frequent guest and good friend throughout the years on WGN. Jim was the horticultural advisor for the University of Illinois Extension Service for nearly 35 years and formed James A. Fizzell and Associates in 1991 to offer his expertise to a variety of entities needing specialized help, including the Chicago Cubs. (Don't blame Jim for the team's performance, he only helps with the grass.)

An entry on Spike O'Dell's WGN webpage.

In September, 2009, at a Ford tractor collectors' show in Union Grove, Wisconsin, I had the thrill of talking with the incomparable Harold Brock, the "Father of the Modern Tractor."

Courtesy: IBA Archives, George Burns

About the only things Oprah and I have in common: the letter O, we both grew up on farms and are both members of the Illinois Broadcasters Hall of Fame. In May of 2011, I gave an introductory speech welcoming her to the HOF. TO MY LEFT: *Dick Biondi, Roger Ebert, Oprah, Bill Kurtis.*

Uncle Bobby tapes the action as I interview a chef at the Peninsula Hotel, Hong Kong, 1974.

Governor Jim Edgar with me at the re-naming ceremony of the Junior Livestock Building at the Illinois State Fair in 1997.

I was given a silly hat to wear during an agricultural tour of England. I don't recall why I was being "honored," but perhaps it was because of my University of Wisconsin shirt.

In May, 2001 I was very flattered to receive an Honorary Doctor of Letters degree from the University of Illinois. With me is Gloria, and longtime friends Joan and Paul Wallem.

Courtesy: University of Illinois

Adding to the Doctor of Letters degree, the University of Illinois honored me in 2012 by choosing me to be a Spring Commencement speaker.

In 1983, Secretary John Block took a trade delegation to the Soviet Union. In this photo, I'm being very careful about what I say about the living and farming conditions I saw because watching Bob Varecha record my comments were some men who I'm sure were KGB.

Fidel Castro outlived several U.S. Presidents who refused to allow American farmers to sell their products to Cuba. In a 1999 trade mission set up by George Ryan, the first U.S. governor to visit Cuba since before the embargo, we found a willing and eager trade partner. Still, in 2012, we wait.

When in Cuba...

In Cuba, 1999. I was either asking a question about Cuban agriculture, or treating our hosts to a "Samuelson Sez" commentary.

"Investigative" reporting from Angelo Papagni's Madera, California vineyard.

2006 in Honolulu, reporting via cellphone during the 65th Commemoration of Pearl Harbor Day.

Dave Marzullo, WGN Radio

In 2010, I was honored to have a Chicago street named after me. Making it more special is that my street intersects with Bob Collins Way (below).
LEFT TO RIGHT: *WGN Radio Vice President and General Manager Tom Langmyer, Chicago 42nd Ward Alderman Brendan Reilly, Gloria, Christine Collins and me.*

Dave Marzullo, WGN Radio

and spiritually."

The United Horsemen are conducting two petition efforts. The first is directed to Congress: "We petition Congress to restore the U.S. horse industry." The second goes to the White House, "We petition the Obama administration to restore humane horse slaughter to improve horse welfare, stop needless and wasteful suffering and create jobs."

I totally support both of these petitions and urge you to do the same, to improve conditions for the abandoned, unwanted horses.

My thoughts on *Samuelson Sez.*

THE ENVIRONMENT AND THE AGENCIES THAT ARE SUPPOSED TO PROTECT IT

Endangered Species Act is EXPENSIVE - November 1, 2009

Ever since the 1970s, America's farmers and ranchers have had to deal with the Endangered Species Act and it hasn't been easy. At times it has been costly to farmers. During the three decades the Act has been enforced, we have shared stories of farmers who were fined and had farm equipment confiscated because they had innocently and unknowingly farmed in areas set aside for endangered species.

Something I hadn't thought about is the cost of the endangered species program; until recently, when *Scientific American* published a new report from the U.S. Fish and Wildlife Service. The 202-page report covered species protected under the Endangered Species Act and listed the money spent in fiscal year 2007. As I learned, protecting endangered species is an expensive proposition that takes a lot of our tax dollars.

According to the report, "The U.S. federal and state governments spent $1,537,283,000 toward conserving threatened and endangered species in 2007, plus another $126,086,000 in land purchases for habitat preservation." The definition of "conservation" is interesting in this report. Conservation includes a wide variety of activities such as "research, census, law enforcement, habitat acquisition and maintenance, propagation, live trapping and transplantation."

The species that required the most money in 2007 was the Chinook salmon. It appears on the list multiple times because it is endangered at multiple sites; $165 million were spent to protect it. Then there was the western population of the endangered Steller Sea Lion; the cost there? $53 million. Gray Wolves, which lost much of their ESA protection this year received only $4.3 million. Receiving more than the Gray Wolf? The Indiana Bat, $6.3 million and the Delta Smelt, $6.6 million.

The *Scientific American* story pointed out that the Fish and Wildlife Service is not a big spender on the ESA, accounting for only 7% of the total outlay. The Army Corps of Engineers spent $212 million and the Federal Highway Association spent $35 million.

In these days of severe federal and state budget problems, with unemployed parents and hurting children, I really question this amount of money being spent on the Endangered Species Act. As a taxpayer, I would much prefer to see many of those dollars spent on programs to help people.

That's my thought on *Samuelson Sez.*

Convention Hallway Talk - January 17, 2010

It is convention time for the many organizations that represent farmers and ranchers, general farm organizations as well as the specific commodity groups that meet during the winter season. I enjoy attending these gatherings, listening to various speakers and interviewing officials and members. But one of my favorite activities is to walk the hallways in the meeting room area and listen to the hallway conversations because that gives me a pretty good feeling of the hot-button issues.

It didn't take very long walking the hallways at the American Farm Bureau Convention in Seattle this month to learn the number one issue in the minds of Farm Bureau members. Time and again I heard the words "global warming," "climate change" and "cap and trade," as well as EPA rules and regulations. Their feelings were summed up well by one farmer who said, "If the world needs food, why don't they let us farm without slapping us with unreasonable rules and regulations. We are good stewards of our environment and were caring for the air, water and soil long before a lot of people even knew how to spell 'conservation' and we are getting better at it every day." Another farmer told me directly, "Keep reminding your audience every week, 'No Farms, No Food'."

Producers were especially riled by the USDA suggestion late last year that eventually U.S. farmers should take 60 millions acres out of corn and soybean production and plant trees to improve air quality. "Ridiculous, that makes no sense at all," were the words used by AFBF President Bob Stallman when I interviewed him. He said with the U.N. calling for a doubling of global food production to feed the growing population, it would be wrong to take the world's most efficient food producers and their acreage out of production. Illinois Farm Bureau President Phil Nelson was applauded by Farm Bureau delegates when he introduced a resolution outlining the negatives of the Cap and Trade proposal in the House legislation, ranging from higher food and energy prices to sending our agricultural production to other countries. It passed by a unanimous voice vote.

Former Texas Democratic Congressman Charlie Stenholm brought hope to Farm Bureau members when he said the House bill in its present form would not make it through the Senate. He went on to say, however, that farmers and ranchers really need to stay in contact with their enators and constantly remind them this is a bad piece of legislation for all Americans. I totally agree and urge you to follow through on Charlie's challenge. No action here means bad action down the road for all producers and consumers.

My thoughts on *Samuelson Sez.*

EPA is Out of Control - October 24, 2010

Last week I talked about issues that you, as agricultural voters, should consider when you go to the polls November 2nd. I mentioned rules being proposed by the EPA, the Environmental Protection Agency. I heard from some of you who wanted to know more about the EPA proposals.

So, I turned to Gary Baise, the top agricultural environmental attorney based in Washington, DC, who spends his time defending farmers and ranchers in environmental lawsuits. Here are the issues that concern him and should concern you.

There is a new solid waste definition scheduled for early next year that will include on-farm incineration units, animal crematories, disposal of animals and agricultural waste.

EPA is coming up with new stationary engine regulations which will regulate existing diesel engines that are larger than 500 horsepower.

EPA is seeking information so it can regulate bioenergy/biogenic emissions to determine greenhouse gas emissions from those sources.

EPA has informed USDA that it is likely to change the standards for ground level ozone because EPA is concerned not only about ozone's effect on people, but also on sensitive trees and plants.

To me, the most dangerous of new rules is the PM-10, or dust standards. EPA is taking into account a new expanded body on thoracic-coarse-particle health evidence which surely suggests a tightening of the dust standard impacting agriculture. Observing the dust clouds surrounding combines in this year's harvest, there is no way you can keep that dust from leaving your property and blowing into a nearby housing development. Yet, I'm sure there are people at EPA who think this can be done.

This is not a partisan fight. Democratic Senator Blanche Lincoln, Chair of the Senate Agriculture Committee, recently expressed her concern about EPA's intrusions such as "unworkable spray-drift pesticide regulations; proposed ambient air quality standards that would impose impossible dust reduction requirements on farmers; wetlands regulations that put even bone-dry areas off-limits to agriculture use and ideological bias toward environmentalists when resolving Clean Water Act lawsuits."

I feel the EPA should not be allowed to put these rules in place without congressional oversight, that obviously is not there now. I hope this answers your questions and I urge you to vote for members of Congress who feel the same way.

My thoughts on *Samuelson Sez*.

Some Interesting Points of View on BLM - February 17, 2011

On *Samuelson Sez*, when I ask for your input on a given subject, boy, do you respond, and for that I am truly grateful. The subject a week ago was rancher relations with the Bureau of Land Management and grazing on public lands. I begin with the shortest e-mail I received, "Please do not refer to the BLM LANDS as PUBLIC LANDS!!! They are FEDERAL LANDS. The term PUBLIC LANDS was attached by the ENVIRONMETALISTS!" That from a federal lands rancher.

But most of the responses sounded like this, from a rancher: "Here in Montana, we have so many issues regarding the management of these lands that it is hard to know where to begin." One of the items listed in that e-mail: "There is a huge push in Montana by some groups to replace the domestic livestock grazing on public lands with bison. To the average Joe living outside the area, this may sound like a grand idea but let me interject some of that local common sense to the issue. People have been domesticating and refining cattle for centuries. One of the biggest reasons for this is to make the livestock as manageable as we can, and at the same time, be very efficient in the way they graze. Centuries worth of work have made domestic cattle far more manageable than a buffalo. An animal that can be easier managed can be easier protected against death and disease and when you can do a much better job of moving the livestock around, or keep them from moving, your land benefits and it benefits even more with a wise local manager in charge."

Space doesn't allow me to share several other serious criticisms of

We spent so much time over the years on the road with our bands doing remote broadcasts, we decided to give it a name.

BLM, but the one mentioned most often is the growing influence of radical environmental groups in BLM decisions and, as one rancher said, "coming from people who don't live in ranch country and don't have to make a living here."

But there was a totally opposite viewpoint from a rancher in Arizona who said,"I agree the agency is out of control and needs to be reeled in but for a different reason. The BLM has bowed to cattlemen's wishes and is removing America's 'National Treasure', the wild horse and burro, from public lands. At the present rate of herd reductions, the existing wild herds will not be sustainable within a matter of years. With less than 2% of U.S. beef cattle production coming from grazing on public lands, perhaps it's time that cattle grazing on public lands should be put to rest. It is as unneeded and antiquated as the no-longer-used cattle drives of a century ago."

Thank you for your input and I close with a personal observation. Before ranchers and cattle arrived on the scene, I wonder who managed the grazing of the hundreds of thousands of buffaloes on the land and who fought the forest fires in the West?

My thoughts on *Samuelson Sez.*

Finally, Some Wins for Farmers - April 15, 2011

It seems to me that every week farmers get attacked or sued by individuals or groups because of environmental practices. They are taken to court, and as one farmer e-mailed me recently, "We get to the courtroom, we don't have the dollars to hire a battery of top lawyers and so we lose the case."

That is why this week I am delighted to share with you two headlines that declare victory for agriculture. Headline number one: "Environmental Protection Agency exempts milk from spill control rules." This is one of those silly rules that the National Milk Producers Federation has been fighting for two years and finally, a few days ago, Lisa Jackson, the Administrator of the EPA said, and I quote, " That exemption is now, today, finished with White House review and will be published today."

This is the proposed rule that tried to put dairy farmers into the same category as oil producers. The Environmental Protection Agency requires operators with oil tanks and containers to write a plan to prevent and control spills. Jackson said the spill control regulation also covered animal fats, which led to the proposal to include dairy farmers with milk storage tanks, since milk contains animal fat. Thanks to this official EPA ruling, dairy farmers don't have to worry. It is a common sense win for agriculture!

The other headline came from the front page of the *Sacramento Bee* in California: "Delta Diversion Plan Takes a Hit When Judge Sides With Farmers." San Joaquin Superior Court Judge John Farrell ruled that the access to farms and property sought by the California State Department of Water Resources amounted to a "taking of land without adequate compensation or protections." The court decision said California state workers cannot set foot on Delta farms to start designing a controversial canal or tunnel to divert water south, the cornerstone of the Bay Delta Conservation Plan. The proposed work would have taken three weeks with trucks, forklifts, and heavy drilling rigs burrowing 200 feet into the soil, and the only compensation would be for any damages incurred.

Thomas Keeling, the attorney for the plaintiffs, said "this was simply a water grab and a violation of farmer property because they were not going to pay farmers for coming onto the land, drilling holes and disrupting planting to carry out the tests they needed." The judge agreed and told the state of California, "No, you can't do that." Another win for farmers and ranchers, and I am delighted!

My thoughts on *Samuelson Sez.*

EPA Administrator Visits Iowa Farms - April 21, 2011

I really don't think there has ever been a strong "friend" relationship between America's farmers and ranchers and the EPA, the Environmental

Horse racing is primarily an agricultural event that I've spent many enjoyable days covering at beautiful Arlington Park. ABOVE: *I'm interviewing a man I've long admired, Richard L. "Dick" Duchossois. His Duchossois Group, Inc., a family-owned company headquartered in Elmhurst, Illinois has part ownership of the corporation that owns Arlington Park and Churchill Downs race tracks.* BELOW: *At Arlington Park, TV weatherman Willard Scott may be giving me the forecast or a betting tip.*

Protection Agency. Farmers and ranchers I talk to about the agency say they feel that it is people sitting in their Washington offices without ever setting foot on a farm or ranch, coming up with ideas on what farmers and ranchers should do for the environment, with little or no understanding of what it takes to produce food.

Last year before the November election, the EPA became a real lightning rod as critics said to members of Congress, "Do something about the rule making mandated by the EPA; the agency is overstepping its authority and taking over the role which should be left to Congress."

Well, something happened a few days ago that I find encouraging and I want to give Secretary of Agriculture Tom Vilsack credit because I am sure this event happened because of him. He traveled to his home state of Iowa and invited EPA Administrator, Lisa Jackson, to accompany him, which she did. They met with farmers and ranchers to discuss EPA's and USDA's joint efforts to ensure that American agriculture continues to be productive. They visited a livestock farm in Pleasantville, a 1600-acre row crop farm in Prairie City and a bio-diesel plant in Newton.

After visiting the farms and spending time talking to the farm families Lisa Jackson made this statement: "These opportunities to talk with farmers on their land and see their operations at work are incredibly valuable. Open communication and transparency are the essential first steps toward protecting air and water quality and ensuring the health of farming communities. Agriculture is part of the foundation of the American economy. EPA's mission to safeguard clean air, clear water and productive land is a critical part of sustaining farming jobs and productivity, and it's vital that we communicate and work together on these issues we share."

Now, some of you may say those are the words she knew people wanted to hear, strictly political talk. But I will give her the benefit of the doubt because I don't think anyone can set foot on a farm or ranch today and not come away with a better understanding of what it takes to produce food and an admiration for the families that accept this daily challenge.

She has the opportunity to prove me right or wrong in July when she will announce the critical rule dealing with controlling dust on farms and ranches. I hope the Iowa farmers told her that containing dust inside their farm boundary is as impossible as catching a falling star. I am glad the EPA Administrator went to Iowa and I thank Secretary Vilsack for making it happen.

My thoughts on *Samuelson Sez.*

THE FOOD AND FARM POLICE

Beware of the Food Police - July 24, 2008

Stand by America! The Food Police could soon be knocking on your door at dinner time to see if you are eating the "right food." Another example of proposed legislation to protect ourselves from ourselves made its way into the news last week.

This is the latest one that caught my attention. Jan Berry, a member of the Los Angeles City Council, is spearheading legislation in that city that would ban new fast food restaurants like McDonald's and KFC from opening in a 32-square mile chunk of the city, including her district. The targeted area is already home to some 400 fast food companies and the reason she's introducing the legislation is to combat obesity which she says is exceptionally high in that area. She legally defines the fast food industry as having characteristics including a "limited menu" and "food served in disposable containers."

Ms. Berry apparently has determined that fast food restaurants are the main cause of obesity and the problem will go away if the law says no more can locate in that area. I find it interesting that she says nothing about individual responsibility in food choice and exercise as being a possible solution to the obesity problem. Just lay it all on fast food restaurants and pass a new law. As a student reading the U.S. Constitution, nowhere did I see a sentence that says that Federal, State or Local governments should decide what we eat, how we eat and when we eat. I certainly recall a line in the Declaration that says we are all created equal, but what we do with our diet and body after that becomes our responsibility, not government's.

Add to that the legislation banning trans-fats and requiring the listing of calories in every restaurant menu item in cities across the country. And then let me add yet another attempt at dictating my diet that I discovered when I visited Denver, Colorado a few days ago. Prior to the Democratic National Convention, a representative of the Democratic National Committee arrived in Denver to tell hotels and restaurants serving food to the delegates and visitors that they were to serve only organic meat and produce, and only locally grown in the Denver area.

That stirred a real fuss in the industry because restaurants and hotels had already ordered their food supplies for the convention. Finally, state officials told her it would be impossible. There simply was not that much locally-grown organic food available, so back off on that demand.

When will we stop passing laws and turning to government to make decisions that we as individuals should make? As my parents always told me, "In this country we have the freedom of choice, make the right one and you benefit, make the wrong one and you pay the price, but whatever you choose, it's you and you alone." It was good advice then and it still is today. Let's stop turning to government to protect ourselves from ourselves.

My thoughts on *Samuelson Sez.*

Whose Responsibility is It? - June 22, 2010

I often wonder, if the authors of our Constitution were alive today, what they would think about the many ways in which the Federal government injects itself into our daily lives and decision-making. For example, should the Federal government tell us what to eat and how to eat it? I wonder if our founding fathers ever intended that would be the role of the Congress and the administration.

Recently, the report of the Dietary Guidelines Advisory Committee was introduced by Rear Admiral Penelope Slade-Sawyer of the Health and Human Services Department and she said, "New nutritional guidelines should focus on keeping Americans from getting even fatter with an emphasis not only on healthy foods but on finding ways to help Americans eat better and exercise more."

I can't argue that there is a reason for that statement with two-thirds of U.S. adults overweight or obese, but what ever happened to personal responsibility? Must we have a federal law on the books to tell us to do what we know we should do?

Let me share some of the other statements in the report: "Americans of all ages consume too few vegetables, fruits, high-fiber whole grains, low-fat milk and milk products and seafood, and they eat too much added sugars, solid fats, refined grains and sodium." The committee's recommendations emphasize "a plant-based diet with plenty of whole grains, fruits and vegetables and moderate amounts of lean meats, poultry and eggs."

It also suggests decreasing sodium intake from current levels to less than 1,500 milligrams daily. There is a call for drinking fewer sugar-sweetened beverages, a call to decrease saturated fat from 10% to 7% of daily calories. "Families need to learn how to cook healthier food and kids need recess," says the report. It suggested development of fish farms to help people afford fish.

All of these proposals will be considered by officials of the Health and Human Services Department and U.S. Department of Agriculture when they draw up new dietary guidelines, something they must do every five years.

Longtime friend and colleague John Almburg and I share an award in the mid-'60s.

Those guidelines will set standards for U.S. school breakfasts, lunches and other federal food programs, so indeed, they could become law. As long as we refuse to accept personal responsibility, we can expect the federal government to encroach further into our daily lives. I don't know about you, but I don't like the idea of "Big Brother" looking over my shoulder in the kitchen or dining room. That is not its job!

My thoughts on *Samuelson Sez.*

Watch Out for the Food Police - October 28, 2011

The Food Police just keep on coming! They are individuals and organizations who want to tell us what to eat, how to eat it and when to eat it, taking away that personal responsibility from us. While America's farmers and ranchers continue to carry the message to U.S. consumers that they have available to them the safest, most nutritious, most economical food in the world, there are a lot of other people out there telling consumers a totally different story.

To make that point, let me share today two recent collegiate events; Harvard University, The Forum at Harvard School of Public Health, looking at agricultural production and food nutrition. Under attack in that discussion was the Department of Agriculture's Supplemental Nutrition Assistance Program, we call it SNAP, it used to be called the Food Stamp Program.

Walter Willett, Chair of Harvard's Department of Nutrition described SNAP as "a conduit for people to buy mostly junk and soda. We're writing checks of billions of dollars a year to buy soda for the SNAP program and with the other hand we're writing checks to treat diabetes. It's nuts!" he said.

Gary Williams, agricultural economist at Texas A & M was on the panel. He challenged the idea that Americans lacked food choice because of the focus on wheat, corn and soy. He said, "I would challenge you to go any place in the world that provides more quality and diversity. Sure, we've got a lot of junk food out there, but you don't have to choose it."

Willett came right back saying, "If we judge by its impact on human health, the American food supply is a disaster. Americans consume huge amounts of refined starch, sugar and red meat and very inadequate quantities of fruits, vegetables, beans, nuts and whole-grain, high-fiber foods."

Then there was the event at South Dakota State University, where the Department of Journalism invited Eric Schlosser, known for his books *Chew on This* and *Fast Food Nation* and his documentary film *Food, Inc.*, filled with what I call outright lies on how we produce and market our food, and he is invited to speak at an Agricultural Land Grant College! But I think agricultural students outnumbered the journalism students in the audience and they challenged Mr. Schlosser on many of his claims to the point where Mr. Schlosser finally just said, "Okay, I will agree to disagree." Thank you SDSU ag students!

My thoughts on *Samuelson Sez.*

CHECKOFF PROGRAMS

Checkoffs are Essential - January 16, 2009

There are several things that America's farmers and ranchers don't really need in 2009. Today I would like to focus on two areas that I think we could well do without in agriculture, in this new year.

First of all, we don't need a family squabble in this relatively small agricultural family in the United States. Secondly, I really don't think we need another farm organization; yet we are getting both in the current controversy over the spending of checkoff dollars in the Soybean Checkoff Program.

Many of you, I'm sure, are aware that the American Soybean

Association has filed an official request with the Office of the Inspector General of the U.S. Department of Agriculture to check allegations of funds being misspent as well as some questionable personal conduct activities in the soybean program. The Association wants them fully investigated and corrected.

Now, along with that we have a group of soybean farmers who have formed yet another soybean organization, the United Soybean Federation. I'm not quite sure why and I certainly don't understand why we need yet another organization in agriculture. Be that as it may, Secretary of Agriculture Ed Schafer, in his news conference at the Farm Bureau Convention in San Antonio last week, said, "the allegations are serious and as the overseer of all checkoff programs, the Department of Agriculture will indeed check out the allegations to determine if they are true or false."

I want to go beyond this single incident however, and talk about the checkoff programs in general, because any "black eye" for one program is going to spill over to all checkoff programs and raise questions about funds being collected and spent. I firmly believe that American agriculture and producers have benefitted from checkoff programs; dollars that have conducted research, promotion, marketing and advertising; dollars that brought us "Beef, it's what's for dinner," "Pork, the other white meat," "Got Milk?," "3-A-Day" and all of the promotions that have helped expand the consumption, as well as the nutrition and safety, of America's agricultural products.

I have been saying for decades that if producers are not willing to invest their time and money into marketing their products, then one of two things will happen. It won't get done at all or it will be done in a way that will not help producers. We need these checkoff programs; we need the staffs and farmer board members to conduct them in a legal and constructive way; and we need to correct this situation quickly before it infects the entire checkoff system.

My thoughts on *Samuelson Sez.*

Are You Inspired? - March 10, 2011

It may be these complicated times in which we live, or it may be the "maturing process," which is my term for getting older, but it seems to me in my life every day, there are more questions than answers. But I am lucky. I have you, my readers, and I can turn to you for enlightenment, direction and advice. This week the subject is advertising slogans, because two of the best known advertising slogans in the world have come from the agricultural community, from beef and pork checkoff dollars:

"Beef, it's what's for dinner" and, "Pork, the other white meat."

But now, after 24 years, the National Pork Board told us a few days ago, "No. We are not going to call it the other white meat anymore; we have a new slogan and we are launching an $11 million ad campaign to target the estimated 82 million Americans who already eat pork." Well, after the success of "Pork, the other white meat" I was anxious to see the new advertising line and here it is: "Pork, Be Inspired." Be inspired for what? Or to do what? I realize change is necessary and I know you can't go with the same line forever, but one pork producer e-mailed me "What about the age-old line 'If it ain't broke, don't fix it?' 'Pork, the other white meat' still works for me."

The National Pork Board has its reasons for searching for new ways to attract consumers to pork. Per capita consumption is slipping, down 2.2 pounds last year to 48 pounds according to USDA, placing it third behind chicken and beef in the U.S. market. Beef sales have also declined, but part of the reason, I'm sure, is the economic downturn and the price advantage that chicken has over pork and beef.

I was surprised that the new campaign is not aimed at attracting new consumers, but the goal is to convince current pork consumers to use more pork in their menus by urging them to be inspired to be more creative with pork recipes. With thousands of new cooks and family households being created every week in this country, it seems to me you must work to attract them to "the other white meat" to increase sales and consumption.

I hope I'm wrong because pork checkoff dollars are precious dollars that must be spent wisely to benefit producer income and consumer satisfaction, but "Pork, Be Inspired" just doesn't do it for me and for me it raises more questions than answers. But, how about you? Does "Pork, Be Inspired" resonate with you as a producer? As a consumer does it encourage you to go out and buy more pork and find new ways to enjoy it? I would like to hear from you. You can e-mail me at orion@agbizweek.com.

My thoughts on *Samuelson Sez.*

AG EXPORTS

Trade is a Two-Way Street - September 10, 2006

At the Farm Progress Show, a week and a half ago, during the hour that Secretary of Agriculture Mike Johanns spent with us on stage in front of a standing room only crowd, a farmer from Illinois stood up to ask a question. Well, he actually made a statement, then he asked a question. He said, "I am certainly in favor of the United States being number one in exports, manufactured goods and certainly in exports of agricultural products, but I don't like the idea of agricultural products being imported into the United States. Mr. Secretary, what can we do to fix that?" That is a challenging and intriguing question. It's a question I have heard before and the Secretary didn't hesitate with his response by saying, "I understand where you are coming from, but then let me quickly say, it doesn't work that way, because trade must go both ways. For us to be able to export our agricultural products or any other items that are manufactured in the United States, we must be willing to take products, including agricultural products, from other countries. Remember there are farmers in many countries around the world who see the United States as a very important market to them and they want access to our market." He went on to say, "We are probably the freest country in the world in allowing products to come into this country without major tariffs and that is one of the reasons we are fighting strongly to level the playing field and open more markets for our products in international trade talks." He concluded, "It must go both ways."

And while many of our cattle producers don't like the idea of beef coming into this country, or while producers of other crops and commodities don't like the idea of foreign products coming in to compete, he said, "if that doesn't happen, then we see borders around the world closed to our products." Also in the crowd that day was former president of the American Farm Bureau Federation, Dean Kleckner. During his term as Farm Bureau president, in many interviews with him, I heard him say several times,"If it doesn't go both ways, it ain't trade!" That may be grammatically incorrect, but the statement is indeed correct. If it doesn't go both ways, it isn't trade. Secretary Johanns pointed out that is why it is important to revive the Doha round of WTO talks, to give America's agricultural producers more access to foreign markets. So he handled the question very well. The questioner probably wasn't satisfied with the answer, but at least he knows now why the Secretary and many people in agriculture stand the way they do on trade being a two-way street.

My thoughts on *Samuelson Sez.*

Another Reason to Thank America's Food Producers - November 26, 2006

Because America's farmers and ranchers do what they do so well,

there aren't too many Americans off the farm who give much thought to agriculture or its contributions beyond the dinner table. Farmers and ranchers provide food, fiber and fuel, but they also provide something else that is very positive for the American economy. I am talking about our nation's "trade balance." Of course, those monthly reports show that for years we have had a negative trade balance. But those same reports also show there is one segment of trade that has a positive trade balance, and the latest report shows it is growing. I am talking about agriculture, the products from our farms and ranches that move into overseas markets.

In its quarterly trade report this month, the USDA stated, "Fiscal 2007 farm exports will surge to a record $77 billion as higher sales of corn, soybeans, wheat and livestock push the country's agricultural surplus to its highest level in three years." To give you an idea of how that compares, USDA boosted its export total by $5 billion in that quarter alone, from the $72 billion estimated in its August report. And in the 2006 fiscal year that ended September 30th, we posted $68.7 billion in exports. So the new projection is an increase of nearly $9 billion over 2006. Now, imports are also moving higher, because we have expanded our taste for foreign food products. Imports of cheese, wine, fresh produce and other products will hit a record $69 billion in 2007, but that still gives us an $8 billion trade surplus in agricultural products. Which countries are the biggest buyers of our agricultural products? Our two NAFTA trading partners top the list. Canada is the #1 importer with shipments expected at $13.2 billion in 2007 followed by Mexico at $11.9 billion. Japan, our top agricultural buyer for decades, now ranks third with $9.1 billion. China ranks 4th, importing $8.5 billion of American agricultural products. So, it's another reason to salute America's farmers and ranchers for their positive contribution to our balance of trade. They aren't able to make it an overall positive balance, yet certainly maintaining and increasing the agricultural trade balance because of increased sales and higher prices.

As I write this the day after Thanksgiving, it's one more reason to say "thanks" to America's food producers and not take them for granted. And it's why I ended my Thanksgiving Day table prayer with, "Bless the hands that prepared this food and bless the hands that produced it."

My thoughts on *Samuelson Sez.*

We Need Trade Agreements - June 3, 2011

In two totally separate conversations this past week, I was asked basically the same question: "Orion, why are you so concerned about finalizing trade agreements with countries like Panama, Columbia and South Korea?"

Those questions reminded me that perhaps many people have never understood or, since this debate has taken so long, have forgotten why trade agreements are important, not just to agricultural producers, but to the economy of the United States. So let me offer a refresher course with some help from Texas Congressman Kevin Brady who recently talked abut the importance of trade and offered some reminders for all of us as we keep working on our members of Congress to make those trade agreements a reality.

The Texas congressman pointed out that 19 of 20 of the world's consumers of food and agricultural products live outside the United States and we need to reach that market on a level playing field. That's right, 95% of our market is outside this country.

Over the past three years farm organizations have seen the agreements get close to passage, only to run into another road block. Most recently President Obama said it was time to finalize the agreements and send them to Congress for approval, and the agricultural community cheered. But then, Nebraska Senator Mike Johanns told me last week, labor unions convinced the Democratic leadership to add a provision to the bill that would sharply increase spending for U.S. worker training programs. He said that put up yet another road block and now the future of the agreements is again in doubt.

In his comments, Congressman Kevin Brady pointed out that the pending agreements with those three countries would increase sales of American products by $13 billion every year. In addition, beyond the benefits for farmers and ranchers, overall economic growth in this country would increase by $10 billion and that would create 250,000 new jobs in this country. The agreement with South Korea would greatly benefit U.S. cattlemen and Sen. Johanns told me that tariffs on U.S. beef would drop from the current 40% to zero over the next 15 years. Australia currently ranks number one in beef exports to Korea, we rank fourth and this agreement could certainly level this playing field.

Now, the other reason it is important to get these agreements approved is our competition, because there are trade agreements, already approved, between the European Union and South Korea as well as other countries that will give them major advantages over agricultural exports from the U.S. Those agreements will become effective in a matter of weeks and officials in Colombia and Panama say they are about to give up because of the lack of action on Capitol Hill.

So, tell your representatives in Congress to get to it. End the debate and amendments; get the trade agreements finalized and benefit all of us in America.

My thoughts on *Samuelson Sez.*

The Value of Agricultural Exports - December 8, 2011

"Why do we ship so much of our food overseas?" That was the opening question in a recent e-mail from a radio listener in Chicago. A lady who has communicated with me three or four times over the past 10 years and always leads with that question, and then follows up with the reason she thinks we ought not be doing that. She says, "If we didn't ship all of that food overseas we would have more here in this country and my food prices would be a lot lower."

Her question this time was prompted by the story a few weeks ago that in the most recent fiscal year we set a record in agricultural exports from

Colleen Callahan was the youngest exhibitor to show a Grand Champion Barrow at the International Livestock Show. Colleen grew up on a farm near Milford, Illnois, studied journalism at the University of Illinois, became farm director for WMBD in Peoria and in 2002, was elected president of the National Association of Farm Broadcasters, the first woman to hold the job.

the United States. We shipped $137 billion worth of agricultural products to foreign buyers, an all time record.

So, after my standard answer of, "How much lower can food prices go?" I expanded my answer to the lady, first of all by pointing out we produce far more than we consume. Secondly, with seven billion people on the planet only 5% live within the borders of the United States, and of the other 95%, millions are faced with malnutrition and starvation. They need what we produce so why should we not share it with them.

There are other reasons, too, why agricultural exports are so important. Let's talk balance of trade; the most recent report showed the U.S. with a negative trade balance of $43 billion for October. The one exception in that report was U.S. agriculture which, month after month for years, has shown a positive trade balance, which means we consistently export more ag products than we import. That's good for the economy, but beyond that the export industry creates millions of jobs in this country. The export of our food also helps build a positive relationship with people of other countries.

I'm sure many of you hear the same question in your community. It is important for those of us in the agricultural community to tell the export story so we can continue to establish new trade agreements with other foreign buyers and increase the global markets for U.S. producers. It's a win-win opportunity for U.S. producers and the U.S. economy

My thoughts on *Samuelson Sez.*

CORPORATE FARMING

The Threat of Monopolies to Agriculture - April 4, 2010

Two weeks ago on *Samuelson Sez,* I talked about the threat of lack of competition and monopolies to farms and ranches in the United States, and I asked for your response. Well you did respond and as promised, this week I share some of your thoughts.

Beginning with a Wisconsin farmer who said, "I was a Monsanto 'seed partner' until a couple of years ago. I've worked with Syngenta traits, too. Been in the seed business over 40 years on a family farm scale. The big seed/trait/chemical companies want all the business. They have a definite monopoly and will continue to do so as long as the legal system lets them."

A dairy farmer in Maryland said, "One of the issues facing the dairy supply-side is the price discovery system. We no longer use the daily Class III values that were derived from the Minnesota-Wisconsin system. That system polled the cheese plants to establish a real-time Class III price. We now establish the Class III price for the whole country based on a very small volume of butter and cheese traded on the CME by a very limited number of traders!"

A farmer in South Carolina shared this: "About 15 years ago, we had four stockyards in a 50-mile radius of our hometown, including one in town. Today there is one left and it is 50 miles away. Yes, I think the hog industry is way too concentrated to the point that farmers are not getting a sustainable price for their animals."

From Illinois, a farmer said, "One of our packer buyers told me the packers used to set the price on beef. Now on Mondays and Tuesdays they fill orders for special programs, then on Wednesday Walmart comes and tells the packers what they will pay. The packers have a competition problem with the large retailers moving the market."

From Missouri: "Yes, it is true that we have a monopoly at work in the seed industry. We have a major company that owns most of our smaller corn and soybean seed companies (about 30) and has agreements and contracts with most of the rest of them, giving them control of about 95% of the genetics. The question as to what to do about this problem is very complicated, but it may take a 'Ma Bell' style breakup to stop this trend."

And finally, from a Georgia farmer: "Agribusinesses have gotten way too big. Because our government has not enforced our anti-trust laws, the farmer has a lack of choices in his seed, fertilizer and chemicals. The monopoly companies are now telling the farmer what to plant and when to plant the crop."

Thank you for your response. But not all producers said lack of competition is a threat. You might be surprised at what some farmers said was an even bigger threat to their well-being than lack of competition. You can read those comments next week.

Again, thanks for sharing on this week's *Samuelson Sez.*

Big vs. Little - Here are Your Thoughts - May 28, 2011

A couple of weeks ago I asked you to share your thoughts with me on the ongoing debate of big farms vs. little farms and once again you responded. Let me share some of your comments.

This one: "It seems that this argument, big vs. little, has been going on forever. When I was in high school in the mid 1960s, I remember our Vo-Ag instructor, Mr. Fred Morris, said there is only one reason to get bigger or make any other management decision and that is to BE MORE EFFICIENT. Of course, most people seem to think that big is always more efficient than little. Not always so."

Another one from a non-farming farm resident who said, "I talk to my farming friends and neighbors and they farm for a living. They don't care if the crops they raise are food, fuel or fiber. If the money outlook is good, the risk is low and it is not too much work, they will grow it. I am not criticizing farmers, but farmers have enough to worry about without worrying about feeding the world."

Then this from a hog, corn and soybean farmer: "I learned a long time ago when I was in leadership positions and traveling a lot not to mention the size of my operation. To some it was huge while others wondered why I bothered to get up in the morning. Many factors contribute to size of operations. For some, being a large operator is their primary goal just so they can say they are. Others, it is strictly a business decision to expand a profitable business. The debate will continue, but economics, management skills and personal preference will dictate the outcome. I suspect the number of producers will decline; there will be closer relationships between producers and their input suppliers as well as the end users of their production."

Then this comment: "I see a need for both small and large operations. My grandparents helped supplement their income with a small five-acre farm. Today I think that would be considered a hobby farm. It is getting harder for today's farmers to remain profitable. If a farmer raises three kids and they want to remain on the farm as adults, then they may need to expand the operation. Bigger operations are needed sometimes, just to break even."

Finally, this comment about farmers: "It does not matter what the size of a parcel of land is that grows food. What matters is that people who grow food care about the land; if they do not, they will not be in business very long. It does not matter if they grow the food organically or not. They are the most honest people I have ever met. Their word is true, their handshake is real; they will do what it takes to feed us and the world, big or little."

And so the debate continues. Several of you took issue with me, claiming my comments showed I was against small farms in favor of big farms. That is not the case. I have said many times over the years there is plenty of room for big farms and small farms, organic farms and conventional farms, and we need them all to feed the world.

Thank you again for allowing me to share your thoughts on this week's *Samuelson Sez.*

FARM SAFETY

Mind Your Manners on Your ATV - October 29, 2007

In the past when I have discussed practicing safety on ATVs, all-terrain vehicles, and particularly when it comes to the safety of young people, I tend to hear from some of you saying "My kids are responsible, they can handle an all-terrain vehicle whether they are six, nine or ten years old."

Well, I still don't agree with that and I will share numbers that support my belief. The American Academy of Pediatrics said recently children under the age of 16 riding these machines is a recipe for tragedy. According to a report by an injury prevention specialist at West Virginia University, the death rate for children on off-road vehicles increased 24% over the past five years.

But now there is another problem and that is protecting the environment. There are more than nine-million off-road vehicles, that's ATVs,

dirt bikes, snowmobiles and other vehicles in the country today and at least one-million new ones are sold every year. Thousands of Americans responsibly use off-road vehicles for work, to explore the back country and to enjoy nature's beauty.

But here is the problem... there is a growing number of riders who are ruining things for everyone by riding off established trails, destroying our public lands and burdening already stressed law enforcement officials. As a matter of fact, the problem is so bad now that the Forest Service called off-road vehicles one of the top four threats to America's forests.

Let me share another number. While off-roaders account for 15% of all visits to U.S. Bureau of Land Management lands, reckless riding represents 50% of all law enforcement incidents, more than thefts, assaults and arson, according to the Bureau.

So it comes back to minding our manners. It is unfortunate that a few irresponsible people are going to spoil it for the majority of people who ride ATVs responsibly; who use all-terrain vehicles for work because they are an important tool, particularly on farms, ranches and construction sites.

But again I emphasize, they are not a toy, particularly for small children. If you think differently, take a look again at the death rate increase noted earlier in this column and I strongly suggest you contact Farm Safety 4 Just Kids and request their booklet on safe use of ATVs.

So, please, follow the rules when you are riding on public (or private) land, and ride safely!

My thoughts on *Samuelson Sez.*

Please Use the Slow-Moving Vehicle Properly - September 6, 2009

At two farm shows I recently attended, I had something very unusual happen, something I don't recall happening in other years at other farm shows. I had seven, count them, seven farmers come up to me and complain about the illegal use of the Slow Moving Vehicle emblem. They said, "Samuelson, you need to talk about it and get something done, so that people know there is a law regarding the use of the Slow Moving Vehicle emblem."

This is something that has frustrated me for years; and again, gets my blood pressure rising. It happens every time I see the SMV emblem used as a driveway marker in towns and rural communities; it makes me even angrier when I see a farmer using the Slow Moving Vehicle emblem on a mail box post to mark the driveway entrance.

Let me refresh your memory about this emblem; triangular in shape, a deep red florescent border and a bright orange center that was designed more than three decades ago by agricultural safety specialists. Its purpose? To attach to the back of tractors, combines, grain wagons and other slow-moving equipment and serve as a warning to motorists on rural roads that they are approaching a much slower moving vehicle in front of them, giving them sufficient time to slow down and avoid a collision.

Because they are easily visible at night with that reflective fluorescent finish, a lot of people now like to use them for driveway markers, to help them find the driveway at night. That is an illegal use of the SMV emblem in nearly every state. Unfortunately, our law enforcement officers don't have the time to ticket everybody using it in that manner. But I would ask that you, if you are using a Slow Moving Vehicle emblem as a driveway marker, please take it down. If any of your neighbors are using it in this way, ask them to please remove it (or put this column in their mailbox and hope they get the hint).

Again, the legal purpose of the SMV emblem is to help motorists avoid a rear-end collision with slow moving farm equipment. Using it in any other way destroys its purpose and meaning, which is to save lives on rural roads and highways. IT IS NOT A DRIVEWAY MARKER!

My thoughts on *Samuelson Sez.*

Safety is Your Responsibility - July 22, 2011

I normally deliver my Farm Safety Sermon twice a year, one at planting time and again at harvest time, two very busy times of the year that can lead to fatigue, carelessness and accidents. Yet I am deeply troubled by what is happening this summer, so here comes a third Farm Safety sermon because it is always dangerous on farms and ranches as has been graphically demonstrated to me by phone calls and e-mails I have received during the past month.

About three weeks ago a farmer friend of mine from Ontario, Canada, called to say, "Orion, in the last four days we lost three farmers in our community. One in an accident on the road, automobile and farm tractor collision; another one, a farmer who went into a grain bin alone without a safety harness, was sucked in and suffocated before emergency crews could reach him; the third, caught in a power take-off that wasn't covered and he died of his injuries. Three farmers in our relatively small community lost in four days."

That was followed a week later by news of a fatal accident in McHenry County, Illinois where I live, about 10 miles from my home. A farmer, well-known in the community was driving his older tractor (without a cab or seat belt) on a state highway and started to make a left turn onto a county highway. At that point, the motorist behind him pulled out to pass, struck the tractor and he was thrown from the tractor and killed.

I know there are times when accidents happen, but look at how these deaths might have been avoided with a covered power take-off, by wearing a safety harness or not going into a grain bin alone, and by the tractor and car driver having a little more patience on the road.

This is a dangerous occupation that seems to be getting more dangerous every year. I can't urge you strongly enough to think "safety" in everything you do. Before you do something that involves risk, think about it, and think beyond to the impact on your spouse and children if they had to spend the rest of their life without a husband or father.

So make sure that all of your safety equipment is working and used properly, make sure the Slow Moving Vehicle sign is plainly visible on the back of your tractors and equipment; do not use it as a driveway marker on your mail box post! And for heaven's sake, think safety with everything you do. I just can't deal with this tragic news anymore.

My thoughts on *Samuelson Sez.*

ESTATE PLANNING

The Future: Let's Look Ahead and Plan for It! - October 22, 2006

While preparing for his first Farm Bill Forum in the studios of RFD-TV in Nashville this past July, Secretary of Agriculture Mike Johanns said one of the topics he wanted to talk about was providing a favorable climate for young people to get into production agriculture today. Indeed that topic did come up at that first Farm Bill Forum, and I think, at every Forum since. There is concern in the agricultural community about the high average age of today's farmers and ranchers. You hear many people lamenting the fact that the only way young people can get into agriculture today is if they inherit the farm from their parents or relatives, and indeed, that is the case in many instances. Yet, there is no sense in wringing our hands over that because it has happened for generations, not just in farming and ranching but in other businesses as well. If inheriting the farm or ranch is how your young people are going to get into agriculture, a word of advice. Make certain you have the succession plan written in legal language so there is no question in anyone's mind what you want done.

It is very difficult emotionally for everyone in the family to sit down with Mom and Dad to talk about life after they are gone, but it needs to be done sooner, rather than later. I have heard too many stories of families being torn apart because language in a will did not clearly spell out the terms of the transition of the farm to the chosen heir. It is equally important to spell out the terms of how siblings who want no part of the farm are able to convert their part of the inheritance into cash. All too often the demand for cash immediately will make it impossible for the person inheriting the farm to have enough resources to continue the farming operation.

There is another way young people can perhaps get into farming today. This idea comes as a result of what I have received over the past year; two letters from elderly farmers with no family and no heirs asking if I could help them locate a young person with a strong interest in farming. Both men, in different parts of the country and not related, said they would be willing to work out a financial arrangement that would bring the young person into the business and to ultimately become the owner of the farm.

I wonder if there is a way we can establish a clearinghouse to bring people like that together; farmers with no heirs who want to leave a farming legacy and young people who truly want to farm, but have no chance of inheriting a family farm. I don't have the answer, but would welcome your ideas on how it might be done because I believe it would be a win-win situation. Some ideas to ponder as we look for the next generation of farmers.

My thoughts on *Samuelson Sez.*

Passing on the Farm - August 26, 2007

For decades we have been talking about the challenge facing a young person who wants to get involved in agriculture and operate, and someday own, a farm or ranch. The assumed rule is if you are not part of a farm family or an heir to the farm, there is no way for a young person to get started in farming today. A recent survey on why dairy farmers left the business produced one response that really troubled me. This was a survey of 333 farmers who participated earlier this year in the Herd Retirement Program of the CWT (Cooperatives Working Together). Of the total, 205 responded to the questionnaire that asked just one question: "Why did you decide to sell the herd and leave dairy farming?" The number one reason: 59% said, "because of the sharply increased cost of production." Other answers included "wanted to retire," "financial difficulties," "environmental pressures" and, the one that really bothered me, 40% of those leaving dairy farming said because "there was no one that could take over the dairy farm." Whether there were children in the family who had no interest or whether it was a couple without children or heirs, they left the dairy business because there was no one there to take over the farm.

This brings me to something we've talked about before. Wouldn't it be nice if we could establish a central meeting point to list young people who truly want to get into farming or ranching on the one hand, and then, on the other hand those farm and ranch owners who have no one to take over the farm when they leave the business. It would be ideal if we could establish a central clearing point to bring those two factions together. That way the people who are retiring and getting out of the farming business would feel good about helping a young person get started in the business and that young person would have an opportunity to fulfill a dream. I know how satisfying that can be because during my years in Chicago, I have worked with three city high school kids who corresponded with me after watching my daily agricultural TV show and expressed a strong interest in becoming a farmer. They were hired for summer jobs by three farmers I knew who had no children, and today, more than 20 years later, two are farming for absentee landlords and one works for a farm implement dealer. I'm not sure how we accomplish this, but I would appreciate any ideas you might have to bring these two factions together and get the job done. Let me hear from

Samuelson Tri-State's Man-Of-The-Year

"U.S. Farm Report" host Orion Samuelson was honored as Tri-State Breeders' Man-Of-The-Year at the cooperative's 42nd annual meeting in Westby.

Samuelson, who is vice president of Continental Broadcasting Company and farm director of Chicago's WGN Radio and TV, was recognized as the "premier spokesman for American agriculture." He travels more than 65,000 miles annually covering national and Midwest farm events.

In his message "Let's Push The Wheelbarrow Right Side Up," Samuelson called the government's 50-cent-per-hundredweight milk-check deduction "one of the dumbest public relations moves ever made in recent legislation."

"The problem is terribly simple, we're not producing too much, we're just underselling it," Samuelson said.

He urged Wisconsin dairy farmers to vote in favor of this month's milk and dairy promotion referendum. Two previous votes on the referendum were defeated.

"Until we see a unified effort on the part of the dairy producer industry in Wisconsin, we're going to undersell," Samuelson said.

Samuelson said customers rank milk third behind coffee and soft drinks. He disputed attacks against the American Dairy Association's promotion efforts by urging dairy farmers to "look at the amount of money they (the ADA) had to spend on market research, promotion and merchandising."

"If you stack the budgets of soft drink companys against that of the American Dairy Association, you're going to find in a tiny little column what the dairy industry has promoted," emphasizing that promotion takes money.

Samuelson called attention to the National Pork Producers Council checkoff that funded the development of "McRib," a pork sandwich now entering the fast food market.

"That's what we call pushing the wheelbarrow right side up," he said smiling. ■

Tri-State's Man-Of-The-Year Orion Samuelson (left) was recognized as the premier spokesman for American agriculture. Director of Public Relations Bruce Odeen made the presentation.

you.

My thoughts on *Samuelson Sez.*

It's an Unfair Tax, Time to Act - September 23, 2010

It's called "The Inheritance Tax," "The Estate Tax," "The Death Tax," and I would add one more title: the "Totally Unfair Tax." Come December 31st, if Congress doesn't take action before then, people who die in 2011 leaving a business, a farm or a ranch will be taxed at more than 50%. That just doesn't seem fair to me; as a matter of fact, it isn't!

A few days ago in the "Wealth Adviser" section of the *Wall Street Journal*, the editors devoted the front page to an article entitled "What Should We Do With the Estate Tax?" They presented two viewpoints. The first was headlined "Get Rid of It. It's Unfair, and There's a Better Way." Ed McCaffery wrote and I quote, "Suppose you overheard a mother schooling her child not to work for regular wages, not to save and, by all means — whatever you do, my dear child! — to spend every cent the little one could accumulate on this Earth and die broke." That is exactly what the U.S. tax system tells the American people. If you work hard, save thriftily and accumulate a fortune, you will be taxed constantly while doing it, and then you will see up to one-half of your savings go to your distant Uncle Sam, instead of the heirs that you choose."

The other viewpoint, expressed under the headline "Keep It. It's Fair and We Need the Revenue," Michael Graetz wrote, "Is it fair for Paris Hilton to inherit her great-grandfather Conrad's fortune without paying any tax on it? Or Yankee owner George Steinbrenner's 13 grandchildren? This is exactly what happens when there is no estate or inheritance tax on the bequests of the very rich. Indeed, that is what the case for the estate tax boils down to: basic fairness. The tax affects a small number of people who inherit large amounts of wealth and who can afford to give up a portion of their wealth to help finance their government."

I totally disagree with that concept. You work all your life to build a farm, a ranch or a small business, giving your blood, sweat, tears and money and paying taxes on it every year. Then when you die and leave it to your heirs, Uncle Sam takes over half of the value. What is fair about that?

We are not talking about the Steinbrenner family or the Hilton family, we are talking about people who don't have that kind of wealth but are still impacted by the inheritance tax. Heirs who would like to continue the business find themselves having to sell part of the property to meet the tax obligation. I have heard from several farmers/ranchers who had to sell so much of the property that it left the business so small it could no longer operate as a profitable farm or ranch. As one farmer told me, "It put me out of business." It's time again to make it clear to your representatives in Washington that it's time to get rid of the Inheritance Tax, to kill the "Death Tax" and do it before the December 31st deadline.

My thoughts on *Samuelson Sez.*

KEEPING RURAL LIVING VIABLE

Main Streets are Here to Stay - August 20, 2008

Guess what! This week I am not going to pick on anybody or anything. As a matter of fact, I am going to pay tribute to some communities in rural America that are thriving and doing well. I know we have seen the stories, three and four times a year, in the *Wall Street Journal* or the *New York Times* that are headlined: "Main Street in Rural America is Dying or Dead; Drying Up and Blowing Away" and yes, that is the case in some rural communities, but not all of them. The inspiration for my comments come from a recent speaking appearance at Greensburg, Indiana, located southeast of Indianapolis. I was there to address the Decatur County Community Foundation Ag Breakfast and heard an interesting story that applies to Indiana but could probably be applied to rural communities in other states as well. A corporate sponsor provided a matching grant fund to establish a community foundation in every county in Indiana back in 1992. The grant was about $2 million matching funds and since then, local businesses as well as private citizens have grown the Decatur County Community Foundation fund to $16 million, all the while using interest from the fund to improve schools, parks, libraries and police and fire departments in the county. It is administered by a Board of Directors, made up of volunteer citizens from the community.

The reason for the ag breakfast was to recognize the importance of agriculture in the county and start a new $10,000 fund in the foundation that will focus on agriculture. As it grows, interest from the new fund will perhaps pay for judging trips for FFA or 4-H teams, or college scholarships for farm kids wanting to further their education. The community spirit shown by the 300 people who attended the breakfast was infectious and real and accounts for the success of this community.

The Indiana visit reminded me of a trip a year ago to southwestern North Dakota to the community of Bowman, a town surrounded by cattle ranches and miles away from a major interstate, all the ingredients of a town that should be drying up and blowing away. But it was doing anything but; the mayor and some of the citizens who took me on a tour of the town of 1,400 people were very proud of their new school, community hospital, museum, senior citizen care center and fire department with training facilities. I was impressed by the fact that nearly every public building was named after a local citizen who had made a substantial financial contribution, again showing pride in the rural community and a belief in its future.

These are just two of many stories I could share with you from rural communities I have visited over the years; towns whose citizens turn aside the purveyors of "doom and gloom" and say, "We're here, we're alive and we're thriving." Hats off to you.

My thoughts on *Samuelson Sez.*

What Makes a Small Town Thrive? - June 9, 2011

Every week it seems, I have more questions than answers. That is why I like to turn to you, from time to time, to get your thoughts and ideas on subjects discussed on *Samuelson Sez.* The subject this week is the "Health

Former Illinois Secretary of State John W. Lewis.

of Small Towns in Rural America."

"Why do some small towns," as one e-mail writer recently put it, "simply dry up and blow away, while other small towns seem to survive and in many cases thrive, in rural America?" It is an interesting question. I have heard many theories why small towns "have dried up and blown away." The interstate highway system gets blamed often because it allows people in the country easy access to shopping centers in bigger cities so they abandon the businesses on Main Street. Then, farms are getting larger, small farms are disappearing, and that leaves fewer people to support Main Street. There is also the Walmart argument. Walmart comes to town and businesses on Main Street give up and close their doors.

That does happen in some rural towns, but as I travel the country I find many exceptions. I can attest personally to the Walmart story because it happened in the city of Viroqua, the county seat where I grew up in

Wisconsin. That community made the front page of the *Wall Street Journal* several years ago because of its success competing with Walmart, simply because Main Street businesses refused to give up.

So, what is the difference? I think it is a combination of pride, spirit, people working together, a strong agricultural economy and a refusal to accept defeat. I saw all of that a week ago when I spent a weekend in West Point, Nebraska, a town of 3,660 people, about a 90-minute drive northwest of Omaha. This year it staged the Nebraska Cattlemen's Ball, a major annual event to raise money to fight cancer that attracted nearly 5,000 people for the day-and-a-half program. The planning started three years ago and when the Executive Committee asked for volunteers to work on the project, more than 800 of the people in the community stepped forward to serve on more than 20 committees. The work paid off with a program that came off without a hitch and raised more than $1 million for the Eppley Cancer Research Center, a record amount for the 13-year history of the Ball.

I saw and felt all of the ingredients I mentioned earlier in the short time I spent in the community with its very proud (and friendly) residents. The citizens of West Point, Nebraska have good reason to be proud of their community!

If you live in a small town that is thriving, let me hear from you at orion@agbizweek.com and tell me why, so I can share it with my readers.

My thoughts on *Samuelson Sez.*

THE CHURCH

Don't Mix Church with Agronomy - December 10, 2010

I always appreciate hearing from my readers and broadcast audience, apprising me of happenings of interest in their area that I would otherwise miss. They very often lead to a topic on *Samuelson Sez* which is the case this week.

I recently received a copy of a story that appeared in the *Fargo Forum* in North Dakota headlined "North Dakota Church Bolts Over ELCA Agricultural Proposal." ELCA stands for the Evangelical Lutheran Church of America and I am a member of a church that is part of that Synod, so it really caught my attention. I know several churches across the country have withdrawn from the ELCA because of the Synod stand on certain social issues, but agriculture? What could the church be saying about agriculture that would offend a rural congregation?

Governor Jim Edgar (1991-1999) was an ardent supporter of Illinois agriculture.

The Synod is drafting a social statement on genetics, and the members of Anselm Lutheran Church in Sheldon, North Dakota in the Red River Valley feel it is an attack on agriculture because it specifically relates to farmers' use of genetic seeds. The ELCA's draft statement says it views plant genetics "with hope and caution," not necessarily because of the science or technology used, but because "the greatest danger in genetic developments lies in the sinful exercise of radically extended human power" that could lead to other sin, such as "exalted pride" or "negligence or complacency."

On an average Sunday, 30 people attend services at this tiny church and when the final count was taken, the vote to withdraw from the ELCA passed 29 to 4. Church Council President Jill Bunn said the congregation felt the church was making statements against farmers, in an area where 95% of the sugar beets are grown using genetically modified seed.

There is agreement in other rural ELCA churches on this issue. A fifth-generation farmer near Jamestown, ND offered these thoughts: "The basic principle I keep coming back to is that I do not believe it is the church's place to give recommendations on farm management practices. We go to church to worship and study Scripture, but from there it is up to individuals to apply the lessons we've learned in our lives."

I totally agree with the thinking of the members of Anselm Lutheran Church and others who wonder why the church takes the time to draft this type of policy. There are many other areas that do lead to "exalted pride" and "negligence or complacency," and I would suggest the 18-member task force that worked on this proposal devote its time to more important issues. And I offer this thought as a member of an ELCA church, the Synod can't afford to lose many more congregations.

My thoughts on *Samuelson Sez.*

More Than Just Christmas Memories - December 23, 2011

For many of us as we mature, the Christmas season becomes a time for remembering, recalling memories of Christmas in your childhood. That's my case. I seem to do it more every year and it takes me back to the hills of western Wisconsin with the focus on the country church, the Brush Creek Lutheran Church, where I grew up; a church with 60 members at the time, that is still in existence today, with about 70 members.

That country church was not only the spiritual center of the community, it was also the social center. It was where we gathered throughout the year for special events and occasions. Christmas time was always a very special time with many memories that I still hold dear, the children's Christmas program where each of us had to memorize what was called "our Christmas piece" and then stand at the front of the church and deliver it to the congregation with the admonition from parents to "speak clearly and loudly so we can hear you." That's one of the memories; it looked like a huge cavern when I was six-years-old and today when I go back, there are just ten rows of pews and it's not very large.

The country church still plays an important role in rural communities across the nation, not just at holiday time but throughout the year. For 30 years on our national television show, I devoted a minute each week to salute a country church, asking members of the congregation to send us history and photos that we could share with our viewers. It turned out to be most popular minute of the show as congregations proudly shared the heritage and history of their church.

As I moved into a large urban area and worshipped at some of the older city churches I began to realize that a country church and an older city church are faced with the same challenge. In the country, young members leave the church and move to a town or city for jobs and a better income; in the city, young people move to the suburbs for the same reason, and in both cases leave the financial support to older members in their declining financial years. The result is many big city churches and rural churches no longer have the support it takes to sustain them and they fade away leaving the community without that important spiritual, social center that we enjoyed while growing up.

So while I remember with a great deal of fondness that church and have the opportunity to occasionally worship there, I would suggest that those of us who have moved away continue to pray for the church and provide financial support so that future generations can have the memories that we cherish today. The Christmas season is a wonderful time to do that. Merry Christmas!

My thoughts on *Samuelson Sez.*

POLITICALLY INCORRECT

Who are the Politically Correct Experts? - October 29, 2006

"Politically correct" is a term that turns me off every time I read it or hear it and I am always curious to know who the experts are who use it as a reason to promote their own views. It is used when discussing politics, diet, physical fitness, ethnic humor, education, world history, environment and a host of other subjects.

A few examples: A school district outlawed playing tag in the school yard because experts said children could be injured. In Wisconsin, recess is being discontinued because the unstructured physical activity of recess should be replaced by organized supervised activity. My reaction: whatever happened to just letting a child's imagination run free? In the play area of the one-room eight-grade country school I attended, I fought many battles between cowboys and Indians (oops, Native Americans) or between U.S. soldiers and World War II opponents that now shall go nameless. Is it necessary that everything a child does today be totally organized in Nike shoes and endless car pools?

The Chicago City Council passed a law that says restaurants in the city can no longer sell foie gras made from the liver of geese or ducks because that is inhumane and now wants to forbid the use of cooking oil containing trans fats. My reaction: since when was government at any level given the right to determine what an individual eats? I see no reference to that in the Constitution and politically correct or not, I will eat what I want to eat.

Agricultural and the environment: I've talked about Iowa wanting to stop the use of manure on soybean fields and Wisconsin wanting to prohibit the spreading of manure on fields between December 1st and April 30th. My reaction: have these "experts" ever farmed and do they have scientific basis for imposing these rules? I get the feeling these are the same people who say big is bad and the small family farm must be preserved, then in the next breath write environmental laws that make it financially impossible for small farms to stay in business.

Then there is the matter of political correctness in race and religion. For example, the insistence by some Native American groups that Chief Illiniwek can no longer be the mascot at the University of Illinois and no athletic team at any level should use Indian or Indian references in their names because it is demeaning baffles me. My reaction: as a Scandinavian of Norwegian descent, I should be incensed by the Minnesota Vikings football team and the silly horn hats they wear, but I guess I'm not smart enough to realize that's an insult to my ethnic background. There are many more examples I will save for another time. Meanwhile, you can totally disagree with everything I have said and call me a politically incorrect nut case. The Constitution guarantees you that right just as it guarantees me the right to think and say what I have just said.

My thoughts on *Samuelson Sez.*

Merry Christmas! - December 8, 2007

My mother didn't have a great deal of higher formal education, but she did have an ingredient that I consider very important, "common sense." One of the words I heard often when I was growing up was the word "respect." My mother kept saying, "Respect your neighbor; respect your neighbor's belongings; respect your neighbor's religion and beliefs. You don't have to agree, but you must respect the fact that there are people with ideas different than our ideas."

I have not forgotten. That is why I am wondering today, what has happened to respect? So many stories that we see on the front page show that respect, maybe like commonsense, seems to be disappearing, particularly during the holiday season, a time when we should respect other ideas and other cultures and other religions.

Instead, we keep getting caught up in being "politically correct," making sure there is no merging of state or religion, trying to be correct in what terms we use in our greetings and our Christmas cards. It gets to the point of being ridiculous; it gets to the point of going into the courtroom once again to show lack of respect for somebody else's religion, ideas or

Courtesy: Pastor John Stemnes-Spidahl

A 2011 Christmas program at Brush Creek Lutheran Church in Wisconsin. I doubt little has changed over the years. These kids were probably just as frightened as I was and they probably received the same warning from their parents as I did from mine to speak their memorized lines clearly and loudly.

beliefs by filing yet another lawsuit.

One of the strangest stories I've heard this holiday season is a major, big-box retailer, who at the corporate level had to decide if they should call it a "holiday tree" or a "Christmas tree." Well, those of us who worship Christ call it a Christmas tree. If you want to call it a Hanukkah bush; if you want to call it a Kwanza tree, if you want to call it whatever term fits your culture or religion, by all means, do it.

Why can't we get along, instead of going into the courtroom to decide whether or not my ideas are more important than yours? It is a time for giving, a time for giving respect to other people and their ideas and not getting caught up in all these silly legal games that seem to come forth under the term "politically correct."

That is why in my case, I will gratefully welcome any greeting you offer during this season and in return, I will say to you, "Merry Christmas" as I enjoy my Christmas tree and Christmas cards.

My thoughts on *Samuelson Sez.*

Be Generous of Spirit - December 13, 2010

It's that time of year when my mail box, and probably yours, is filled with Christmas cards with the theme "Peace on Earth and Good Will to Men," or "Peace on Earth to Men of Good Will." Yes, that should be the spirit of Christmas, but why is it so difficult for some people to allow that to happen? Why can't I say "Merry Christmas" to a person on a Chicago street without getting a strange or nasty look? Why can't people of different religions, different cultures and different countries respect each other's religion and beliefs? Why can't we get along?

Again this year I am seeing headlines saying this group is filing suit to keep a Nativity scene out of a town square, or another lawsuit to keep the symbols of Hanukkah or other religions from being displayed alongside the Christ Child. More often than not, the American Civil Liberties Union is involved and of course, the only people who make money are the attorneys who go to court. But even worse than the lost time and money is the discord that occurs in a community, putting neighbor against neighbor. Why can't we get along?

It is a great time of year to learn about the religions and cultures of other people. Thirty years ago my wife and I moved into a new home on a street of eight homes in a Chicago suburb and soon learned we were the only Christians on the street. Over the years our Jewish neighbors invited us to their religious holiday meals and events and we did the same. I learned so much, but the greatest lesson was that while we may take different paths, our goals are the same: peace and love. We can get along!

Chris Krug, the editor of the *Northwest Herald*, our daily newspaper in northern Illinois, recently shared his thoughts on this subject. "Merry Christmas. These are just words, and not unkind words. Not words intended to hurt anyone's feelings. They are expressions of joy; not pain, not humiliation. They are sincere, perhaps some of the most sincere words that many of us speak all year long. Christmas is a beautiful time of the year for so many Americans, a time when there not only is much to share but also an uncommon willingness to share, a time of peace for all mankind."

So, let's get along and put aside our differences, let our religious displays stand side by side and tell the American Civil Liberties Union to take a rest and enjoy their holiday. Hopefully, the end result will be "Peace on Earth and Good Will to Men," and wouldn't it be wonderful if that could last 52 weeks a year.

My thoughts on *Samuelson Sez*

POLITICS AND GOVERNMENT

My Suggestions for Change - September 28, 2008

Again this week my focus is on Capitol Hill, with a strong message to the members of the House and Senate. It is time to stop the "blame game," to start some serious work on fixing the challenges that currently face our economy and then come up with solutions that will last more than a year.

Many of us have been watching the congressional hearings on Capitol Hill and in my case, wondering about the economic knowledge of the people who make our laws. I appreciated the statement by Don Bright, a member of Bright Trading in Chicago, who told the *Wall Street Journal*, after watching the congressional hearings; "I am watching these gentlemen on television talking about something they have no clue about." Indeed, some of the questions and statements being made by our legislators certainly fortify what Mr. Bright said.

Again, my question is the same as it was a week ago. Our presidential candidates have been serving in the Senate for anywhere from four years to nearly three decades. Where have they been the last two or three years, because we have seen the situation developing for at least that long; beginning with the crisis in the housing industry, followed by the auto industry and then the banking and credit industry. This crisis didn't just happen a week or a month ago. So why didn't Congress, including our presidential candidates, act earlier? Why did they wait to the point where the market was, according to Warren Buffett "nearly totally dysfunctional"?

I think one of our problems in business and society today is that we have lost most of the Depression Era generation, people who went through the Depression of the '30s and learned a hard never-to-be-forgotten lesson. I vividly remember as a young man on our Wisconsin dairy farm, hearing my father say over and over again, "Don't buy what you can't afford and don't borrow money you can't pay back." That common sense economic rule has been ignored by lenders, borrowers and legislators and now we all pay the price. That mortgage offering that sounded too good to be true, indeed, was too good to be true!

Now let me really go off the deep end. To give our legislators the time to do what we pay them to do, legislate, I am going to suggest a couple of changes to our political world. I think we should increase terms in the House to four years instead of two, so House members have some time to take care of business and don't have to start running for office the day after the election. Finally, I think presidential campaigns should be limited to six months. It would save money as well as saving us three years of the same speeches, and it would keep any members of Congress running for president in DC much longer so they could do what we pay them to do.

My thoughts on *Samuelson Sez.*

Where the USDA Budget Goes - February 28, 2009

Throughout the course of the year I get letters and e-mails from listeners, most of them not directly involved in agriculture, who are worried about or complaining about the size of the budget of the U.S. Department of Agriculture. One recent e-mail from a listener said, "With only two percent of the people in this country involved directly in farming or ranching, why do we spend billions of dollars in the Department of Agriculture for programs that really don't benefit most of us?"

I would like to set the record straight on that budget this week and it's timely, because a few days ago the House of Representatives approved the USDA budget for the upcoming fiscal year total $108 billion. That is a lot of money, but when you put it alongside Bank Rescue, Stimulus, Automobile Rescue, it isn't all that large.

Now let's look at where those dollars go. The lion's share of the USDA budget, 70%, totaling $75 billion, $660 million goes to domestic food aid programs. WIC, Women, Infant and Children's Nutrition; another program for child nutrition and then, of course, the biggest one of all, Food Stamps. Incidentally, it is no longer known as the Food Stamp Program in the new budget; it is now the Supplemental Nutrition Assistance Program, (I guess we call it SNAP) and that goes to benefit the more than 31 million people who are now on the Food Stamp or SNAP Program. Again, $75 billion of the $108 billion goes to non-farm programs.

But it doesn't stop there; another $2.5 billion goes to foreign food aid to help feed people in developing countries. Other programs that benefit all of us beyond just farmers and ranchers include Food Safety and Meat Inspection... $3.5 billion; environmental programs that include the various conservation programs to preserve our vital natural resources account for about $2.3 billion and we can't forget funding for the Forest Service Program.

So, what about the area that concerns most of the letter-writers and e-mailers? Subsidy programs and direct payments to farmers. Those programs this year amounted to about $7 billion, about 10% of the amount going for food programs in the USDA budget. Note, also, that the proposed stimulus program will add another $10 billion to the Food Stamp or SNAP program.

I'm not complaining about that, but I just want to set the record straight on where the dollars go and remind everyone that only a small percentage goes into producers' pockets.

My thoughts on *Samuelson Sez.*

Congress, It's Time to Act Responsibly - May 30, 2009

There is a great deal of attention being focused on the legislative agenda of the Obama administration and the many new proposals that President Obama is sending to Capitol Hill. I'd like to remind members of Congress this week that there is some old business on the books that needs their attention and action in this session.

For example, there is WRDA, the Water Resource Development Act; we have been waiting for well over a decade for funding and work to begin on renovating the locks and dams on the critically important Mississippi River system. There is general agreement, even on Capitol Hill, that this work is long overdue, but we are still waiting and it's time for Congress to make it happen.

The other piece of legislation that is important to the nation and, especially to agriculture, is immigration reform that would make it easier for migrant workers to come across the border and do the work on many farms, ranches and dairy farms in the United States that American workers simply won't do.

I'm a little tired of hearing people say that migrant workers take jobs away from American workers. Let me share one example: Several years ago a friend of mine in Michigan, who has a large orchard, decided he would do something to help the unemployed and the unemployment situation. He chartered a bus, sent it to Chicago to the employment office to get unemployed people looking for day jobs to come to his orchard to pick apples. They managed to get about 30 people to board the bus early in the morning, drove to Michigan, and the workers were put to work hand-picking apples in the orchard. By 10 a.m. they were complaining, saying "This work is too hard, we are not going to do this anymore." By noon the bus was loaded and on its way back to Chicago.

No, when it comes to picking strawberries or lettuce in the fields in California, to hand-harvesting fruits and vegetables, or the "grunt-work" of landscaping; many of us, certainly myself included, really don't want to do that hard work anymore. That's where the role of the migrant worker becomes critically important to farmers and ranchers.

But under current law, it is increasingly difficult to meet all the legal demands for workers and employers to make sure crops are harvested and cows are milked. This is not easy legislation because it is a very emotional issue and Congress didn't have the guts to act on an immigration reform bill in 2008, an election year. We need to remind our legislative leaders again that this is important and it's time they show some backbone and get it done. We have waited too long.

My thoughts on *Samuelson Sez.*

Stop the Games, Get Down to Business - July 30, 2011

The purpose of this commentary is not to totally condemn the President of the United States or every member of Congress because over the years, as a media person, I have interviewed many of them and found most of them to be honest, hard working people who have their ideas of how the business of the country should be conducted.

But there are times when I question their judgement and their qualifications to run the business of the United States and this is one of those times. The recent debate over the debt ceiling is a prime example. I don't know about you but I am so tired of seeing the Speaker of the House and the President of the United States appearing on television on an almost daily basis to, as they say, explain why they are doing what they are doing, but instead attacking the opposition and not offering any ideas or compromise to find a solution.

Whether it is Republican or Democrat, that is the direction it has been going in this ongoing debate that would be almost laughable if it weren't so serious. One thing that comes through loud and clear to me is, very few members of Congress have ever owned or operated a business or a farm because if they did, they would have a better understanding of economics. They would know you cannot spend more money than the business generates and stay in business very long. But in Congress, the thinking is totally different. If Congress spends more money than taxes generate, it simply raises the debt ceiling and prints more money. It has been doing that for a decade and that is why our debt is at a level we can't even comprehend. But we do know that eventually we, and our children and grandchildren, will have to pay the bill.

Another personal observation: candidates generally campaign on a platform of change to improve the well-being of the country and its citizens, but when elected their mind-set seems to change. They decide they want to make a life-long career in Congress and will do whatever it takes to reassure

On stage with Ag Secretary Tom Vilsack, Max and our band Roundhouse in the air conditioned comfort of the ADM tent at the 2012 Farm Progress Show in Boone, IA.

re-election. Maybe it is time for term limits!

I do know it is time to stop the game-playing on Capitol Hill before the uncertainty throws the global economy into another tailspin. During the August recess when members return to the home district to visit county and state fairs and hometown festivals, take the time to tell them you have had enough of their childish antics in the halls of Congress.

My thoughts on *Samuelson Sez*

Which Ox to Gore - November 18, 2011

A Congressman recently told me that they need traffic police in the hallways of the House and Senate Office Buildings on Capitol Hill because they are packed with lobbyists going from one office to the next to make sure their programs are not cut or sacrificed in the debt-reduction debate going on in the Super Committee.

We all remember this past summer when we avoided a financial calamity by forming what is called a Super Committee to come up with a debt-reduction program of well over a trillion dollars by November 23rd.

Well, here we are. The Super Committee has been working hard according to all reports and it is probably the least popular job in Washington right now because they are getting hammered by lobbyists from every segment of the economy and the tax-paying public, and that includes agricultural groups.

I have yet to meet anyone who doesn't feel we need to cut government spending; they all say, "Yes, severe cuts need to be made." On the other hand I have yet to talk to anyone who will say, "Cut my program." With every person, with every group, the line is, "Gore someone else's ox, don't gore mine because my program is critically important."

Even though the Department of Agriculture budget is about 1% of the federal budget, the lobbying for USDA programs has been intense. Nutrition programs account for nearly two-thirds of the budget, but don't cut those programs. Lobbyists for soil and water conservation, agricultural research, a farm safety net, export promotion and food safety all say the same thing: these programs are critical to the well-being of the country.

So, the question begs, "Whose ox will be gored?" I can't imagine the pressure on the members of the Super Committee as well as all members of Congress right now as lobbyists and their constituents apply the pressure.

But if we are truly serious about avoiding a repeat of the volatile world markets we saw last summer and the current global impact of the Euro Zone debt crisis, then we must insist that our government does what most of us have to do, live within a budget. That means sacrifice on the part of all of us and everyone's ox needs to be gored.

My thoughts on *Samuelson Sez.*

THE BANKING INDUSTRY

History is a teacher - August 17, 2007

History can and should be a good teacher. In school I always enjoyed the subject because I felt there were lessons to be learned from those who came before us, lessons that perhaps could help us avoid mistakes. Unfortunately, our memories are short and unfortunately, all too often lessons learned, quickly became lessons forgotten. We have a "prime" example right now in the crisis in the sub-prime and home mortgage industry. You should recall several years ago when the savings and loan debacle cost companies billions of dollars, many people lost their jobs, homes and investments, and billions of government dollars were spent to bail out the floundering savings and loan industry.

It all happened because of questionable lending practices on the part of the industry. It's apparent that lesson has been forgotten, so here we are today in a crisis that is shaking the economy around the world; not just in the housing industry and not just on Wall Street, but we have watched the agricultural commodity markets experience some down-limit moves because of the crisis. And yet, had we paid attention to what happened in the past, perhaps we could have avoided what is happening now. Just reflect on some of the commercials you have seen offering credit, whether it be for the purchase of automobiles or houses; commercials that offer loans without a credit check, with no bank statement or income tax returns, no closing costs and with interest only and no principal payments for the first two years. Attractive indeed, and it enabled many people to buy homes they otherwise couldn't have afforded and, in light of today's crisis, can no longer afford. There is plenty of blame to go around here... the lenders for these borrowing practices, the borrowers for not being realistic about their credit situation, the government agencies responsible for oversight of the industry and not enforcing stricter rules for lenders and borrowers.

The lessons were there had we remembered the history of just a few years ago. History is full of lessons of how we as investors should deal with

With the Calvin Hartter family of Eureka, Illinois, winners of a new IH combine.

the ups and downs of the market place. I would close with these history lessons I have learned... prices never go the same way forever, there is always a correction and, as former Secretary of Agriculture Earl Butz told me many times "the cure for high prices is higher prices and the cure for low prices is lower prices." Let history be our teacher.

My thoughts on *Samuelson Sez.*

A Lesson to be Learned - October 19, 2008

With all of the stress and hand-wringing over Wall Street and the global economy over the past month, it is difficult to find something positive to talk about, but let me try.

This week I turn to a longtime friend, whose judgement I truly respect, Porter Martin, President of Martin, Goodrich and Waddell in northern Illinois, a firm that deals in farm management and farm real estate. Four times a year, Porter puts his thoughts in writing in the company newsletter entitled *Seasons.* His most recent effort was headlined "Why Farmland is Your Safest Harbor in the Current Financial Storm." In addition to talking about that though, he offers some history on how we get into the problems we do when government regulates every detail of private enterprise and tries to solve every economic problem with credit expansion, subsidies and restrictions.

Quoting Porter, "Mercantilism birthed and built the two mortgage giants which became the rotten foundation of U.S. mortgage debt, Fannie Mae and Freddie Mac, created by a Congress eager to expand home ownership for lower income people. Fannie and Freddie underwrote half of the $12 trillion in U.S. home mortgages in America, many of them by buying bundles of mortgage-backed securities. When home prices fell, mortgage defaults rose and those securities crashed, throwing the entire banking system into chaos."

Fast forward to the latter part of Porter's column. He said "I hope administrators of any future Wall Street 'rescue' will look back to the farmland price crash of 1981 to 1986, and apply its useful lessons on the need to find real values in the market place, so real recovery can flow again. Lenders struggled through the 1981-86 farm defaults on their own. The Farm Credit System consolidated. Contract sellers took back farms. Banks resold foreclosed farms. Land values in many regions settled and stabilized at new lows, often 60% down from the 1981 peak, and new buyers got a toe-hold in land. And one of the most enduring legacies of that land bust; farmland borrowers and lenders have remembered ever since that leverage cuts more deeply in a price down-trend than it enhances returns in an up-trend. As a result, we do not see over-leveraged farmland deals today. Many sales are all cash, no financing."

It's too bad we quickly forget that history can be a harsh teacher and we pay dearly for not paying attention. Let's hope this year's economic lessons will not be quickly forgotten. Thank you, Porter Martin, for your common-sense thoughts.

I'm pleased to share them on *Samuelson Sez.*

Do You Have the Answers? - February 6, 2009

I have always felt there is no such thing as a dumb question. There are only dumb people who will not ask the question to learn more about the subject and to get an answer. So today, a couple of questions about the current economy.

Question #1. How can so many, well-paid, supposedly intelligent business people in the banking community, not only in the United States, but all around the planet; how can so many people all make the wrong decisions at the same time? But as we look at the chaos in the financial community around the world today, it has become very obvious that all of these well trained and experienced financial people made some very stupid decisions. I understand many of the choices made were based on greed; but I've always maintained that capitalism is based on greed. But, it only works when greed is controlled and it fails when it becomes uncontrolled. So because the world-wide major banks made the wrong decisions, now we are all suffering and paying for it.

There is an interesting footnote to this situation; most community and rural banks are in much better shape than their big-city cousins because they continued their sound lending practices. I'm sure many rural bankers learned and remembered the hard lessons of the '80s.

Question #2. With all of the discussion and debate on the economic stimulus plan, I find it interesting to see governors and mayors from all over the country rushing off to Washington, hopefully none of them are flying in corporate jets; but rushing off to Washington to ask for their share of the bail-out money.

Isn't that admitting on their part they have not done the job they were elected to do? It's obvious they should have been doing a better job of fiscal responsibility managing the finances of the city or the state. But now, with the smell of free money in the air, they run off to Washington to ask for part of that money, admitting to themselves, I hope, "we did a bad job,

To my left, my son David, daughter Kathryn, Lura Lee Ryan and, behind her, Gloria, listening as Governor George Ryan spoke at a party in my honor in 1998.

so now we're coming here to get our share of the money."

And then there's the third question: Where does the money come from? We know the answer to that one; it comes from your pocket and mine as taxpayers because sooner or later, we and our children and grandchildren, and probably, our great grandchildren, will pay the bill for the "rescue plan."

Those are my questions. If you have some answers, I'd love to hear from you.

My thoughts on *Samuelson Sez.*

USDA

The Secretary Sez - Dec. 31, 2007

This is a tradition that is now in its 31st year; at the end of the year my *Samuelson Sez* becomes *The Secretary Sez*, this year the thoughts of Acting Secretary of Agriculture Chuck Conner:

"Well Orion, let me just say that this is a tremendous time to be Secretary of Agriculture and I'm honored and humbled in many ways to serve in this capacity.

"We have seen some remarkable events over this past year with producers responding to the challenge to produce product not only for our food needs in this country, but also now for a substantial part of our energy needs as well. I have great appreciation for what a lot of our livestock producers are going through with higher feed prices, but at the same time I am quick to point out that we are actually feeding more corn for feed today than we were two years ago. So, despite the tremendous increase in demand, our producers are responding to this.

"My natural instinct is to always let producers manage the market; they do a lot better job of managing the market than the Secretary of Agriculture or USDA. If you turn them loose and give them the right price incentive, they will respond in a big way. We are excited about what is happening in rural America.

"I'm also excited that in addition to growing more and more of our energy needs, we are exporting more and more of our agricultural products all over the globe. Dairy products are just one example. We didn't export dairy products five years ago, but today 13% to 15% of our dairy products are exported all over the globe because consumers in other countries discovered American cheese and love it. These are great times for American agriculture around the world.

"Our challenge in the future... we want to make sure these times are not a 'boom and bust' period where we are going to see things fall off the cliff into recession. A lot of economists still predict a 'boom and bust' scenario for today's farm economy. Our challenge is to make sure that we have a good farm bill in place, a good trade policy in place so that this is not a 'bust' later, that these prices are sustainable through a whole generation of farmers. There is nothing I would like to see more than the opportunity for our producers over this period to have good economic times. They deserve it because they do so much for all the American people and I think they should be rewarded accordingly."

The thoughts of Acting Secretary of Agriculture Chuck Conner shared with me in his office at the end of 2007.

That's this year's *Secretary Sez.*

Congratulations, Tom Vilsack - December 20, 2008

Let me begin by offering my congratulations to Tom Vilsack on being appointed Secretary of Agriculture by President-elect Obama. That's three in a row for governors in that position. Mike Johanns was the Governor of Nebraska, Ed Schafer served as Governor of North Dakota and Tom Vilsack served as Governor of Iowa from 1998 to 2006.

At the same time though, I might also offer my sympathy because Mr. Vilsack is facing a challenging job. I vividly remember back in 1980 when Ronald Reagan was elected President and, as he looked over a list of eligible candidates for Secretary of Agriculture, my name was on that list;

I might say far down the list, however. When I learned of it, I called my friend, former Secretary of Agriculture Earl Butz and said "Earl, what do you think?" "Well," he said "Let me tell you; if you become Secretary, 50% of the people in the country will love you, 50% will hate you. John Block was named Secretary, but I have never forgotten Earl's words.

It didn't take long for the detractors to make their objections to Vilsack's appointment public. Andrew Leonard, who writes the *How the World Works* column, stated his feelings in newspapers across the country the day after the appointment was announced. Under the headline "VILSACK WILL BE BIG AGRICULTURE'S MAN IN THE WHITE HOUSE," he wrote: "Barack Obama's nomination of former Iowa Governor Tom Vilsack for Secretary of Agriculture poses an interesting challenge to food policy progressives and environmentalists. It's likely that some of the same people who applauded the nomination of Nobel Prize-winning physicist Steven Chu as Secretary of Energy because it signaled a return of respect for science in the White House, will be disappointed with Vilsack because of his own fondness for science — that is biotechnology. Make no mistake, the biotechnology industry and big agribusiness corporations are mighty pleased with the prospect of Vilsack as ag secretary." And that is just the start, as Tom Vilsack takes over the federal agency and its budget that was born in the Abraham Lincoln administration.

Speaking of that budget, I think it's important for the new Secretary to constantly remind the critics of "big spending" on agriculture that two-thirds of the annual USDA budget goes to provide food assistance programs for people in need, leaving just one-third for all of the programs from research to forest service to conservation and everything in between for the people in production agriculture. These are the people who put food on our table, clothes on our back, a roof over our heads and now, energy in our gas tank. Let us never forget that.

My thoughts on *Samuelson Sez.*

The Secretary of Agriculture Sez - December 28, 2008

As we come to the end of another year I continue a tradition that has been a part of my schedule every year now for more than 30 years. In December, I travel to Washington, DC, sit down in the office of the Secretary of Agriculture, and talk to the Secretary about significant events in the year just ending and look ahead to the new year.

Part of that visit includes *Samuelson Sez,* except this one time a year I [illegible] it *The Secretary Sez*. A few days ago I was in the office with Ed Schafer, former Governor of North Dakota who was sworn in as Secretary January 28, 2008 and who will leave the office January 20, 2009. Toward the end of our visit, I said to Secretary Schafer, "Okay, it's your turn, *The Secretary Sez.*"

This was his response. "Well, one of the things that has been so important to me here is to learn the strength of agriculture. Heading up the United States Department of Agriculture reminds me when Abraham Lincoln founded the USDA, he said it's the peoples department. He called it the peoples department because it affects so many peoples lives in so many different ways. As I look at that from this perch of Secretary of Agriculture, I understand the strength that agriculture brings to this country.

"As we now see a renewal or revival in agriculture as things are moving in, more importantly the energy field, we see that strength come back, not only in the economy but what agriculture delivers as human beings to the foundation of the United States; a hard work ethic, an honesty, a courage; a lending of hand to a neighbor; lend a helping hand to someone in need. Those kinds of things are the foundation of this country. It comes from agriculture; it comes from working the land. I am so pleased to be able to understand and feel that strength of agriculture, because I know it is the foundation of what made this country great."

The *Secretary Sez*, the thoughts of Secretary of Agriculture Ed Schafer as he prepares to leave that office and move back to his home in North Dakota.

One of my fond memories of the Secretary will be sitting on the outdoor stage with him at the Farm Progress Show in Boone, Iowa last August listening to him respond to questions from the audience, handling each one with grace and intelligence. Personally, I would echo what the Secretary said. It is, indeed, a privilege for me 365 days a year to serve the most important minority in the world: YOU, the American farmer and rancher. Thank you for letting me be a part of your life every week.

My thoughts on *Samuelson Sez.*

The Secretary Sez - December 31, 2009

Once a year, the title of this commentary changes from *Samuelson Sez to The Secretary Sez.* At year's end we ask the Secretary of Agriculture, this year Tom Vilsack, to share his thoughts with America's farmers and ranchers.

So picture us sitting in the office of the Secretary of Agriculture about two weeks before Christmas talking about issues of the year ending and a look ahead to 2010. Then I turn to Secretary Tom Vilsack and say "Time now for *The Secretary Sez*" and he responds:

Governor "Big Jim" Thompson and, from left, his wife Jayne, Gloria and me.

"I really appreciate this opportunity to talk directly to farmers and ranchers around the country. Recently I had the sad responsibility of doing a eulogy for Norman Borlaug who passed away last year, The Father of the Green Revolution. During the course of the eulogy preparation, I began to research some of our agricultural history. I think a lot of Americans forget that it was not too long ago that we were a country not capable of producing food for our own people, much less for the rest of the world. But our farmers and ranchers took risks; they embraced new technology, a new knowledge; they looked for ways to become more efficient. Over the course of Norman Borlaug's life, we went from a country that couldn't make ends meet from a food perspective to a country that was certainly the breadbasket of the world.

"So first and foremost I want to thank every single farmer and rancher who is watching the show today and those who produce our food, our fiber and our fuel, for the extraordinary work you do in meeting the needs of American families.

"I also want to thank you for your contribution to the American economy. You know that as a result of your efforts we have a trade surplus agriculture which creates wealth. But we need to focus on a new rural economy to help you stay on the farm. That is why USDA is going to be actively engaged in expanding broadband access, making sure that the energy titles of the farm bill are aggressively promoted; working on climate legislation to bring those additional dollars and offsets into rural America, and making sure that there is a connection between local production and local food.

"So I want to wish you a Happy New Year and again thank you so much for all that you do for us."

Sitting in the office of Secretary of Agriculture Tom Vilsack in front of a television camera and a radio microphone, and sharing with you, not *Samuelson Sez*, but the thoughts of the Secretary on *The Secretary Sez*.

And to all of you my sincere wish for a happy and healthy 2010 and beginning next week, it will be another year of inflicting my personal biases, prejudices and opinions on you every week on *Samuelson Sez*.

Secretary Sez - January 10, 2011

Every week in *Samuelson Sez* I express my personal opinions, prejudices and biases, not expecting agreement, but hoping to stimulate thought and discussion on the subject. The one exception comes at the end of the year when I sit down in the office of the Secretary of Agriculture and re-title the segment *Secretary Sez*. I now share this year's thoughts from Secretary Tom Vilsack:

"Well, I just want to extend a Happy New Year to the 900,000 folks I talked about earlier; the farm families of this country, the folks who produce our food, our fiber and our fuel. I think they are one of the most underappreciated segments in America. These are folks who give every one of us, as consumers, a tremendous advantage. We go into our grocery stores and we leave them with more of our paycheck in our pocket than anywhere else in the world.

"That allows us to buy better cars and bigger homes and take vacations and put money aside for our child's college education or our retirement. We need to thank farmers for the advantages that they provide through their hard work, their effort, the risks that they take. We also need to understand and appreciate that we have the greatest abundance, the most fabulous choice of food products of any people in the world.

"My heart goes out to those folks every day who wake up very early in the morning, who put themselves at risk, who put their financial well-being at risk every single day, to make sure that my family is well fed. I hope the folks who are reading this today who are not farmers, will take time through 2011, just to thank a farmer, for providing us with the food, the fiber, the fuel and a little extra cash."

I then asked Secretary Vilsack the question that I have asked every Secretary for the 35 years I've been doing these year-end interviews: Are you still having fun? He responded, "This is the second time this week I have been asked that question, the last time was by the President of the United States. He asked me if I was having a good time and if I enjoyed this job. I said, 'Mr. President, I have the best job in America.' Then I realized maybe that was not the right thing to say to the President, so I said maybe you have the best job. He said, 'Oh, no, no, no. You may have the best job.' So, this is an extraordinary honor. Every day I wake up thanking God, thanking the President, thanking this country for this opportunity. Sixty years ago I started life in a Catholic orphanage and on my 60th birthday I was in the White House where the President and Vice-President were singing Happy Birthday to me. Is this a great country or what?"

Those were the thoughts Secretary of Agriculture Tom Vilsack shared with me in his office just a few days before Christmas. And I thank him for this annual presentation of the *Secretary Sez*.

Let's Work on the 2012 Farm Bill - June 25, 2011

We are halfway through 2011 and 2012 will be here before you know it. That is important to America's farmers and ranchers because 2012 is the year we are supposed to write a new farm bill. Already, some folks are saying Congress will not be ready to complete a bill that soon and we will probably have to extend the current bill until Congress can do its work.

I think it is a little early to make that assumption because already both the Senate and the House Agriculture Committees are holding hearings around the country and Secretary of Agriculture Tom Vilsack is certainly talking about the new farm bill. Nearly all of those discussions center on budget cuts. Secretary Vilsack recently told producers to be prepared to accept the fact there will not be as many dollars for agriculture in the new farm bill as there are in the current bill. So the lobbying begins on whose programs remain intact, and which programs see the budget knife.

Interestingly enough, and this is no surprise because it's human nature, every government agency feels its work and its programs are critically important, so "gore somebody else's ox, not mine." Soil and water conservation... important; nutrition programs, terribly important to a lot people; forest service, important; food inspection, export promotion, all se programs in the Department of Agriculture are important to those

At the 1997 dedication of the Junior Livestock Building with members of the Eddie Vodicka Band, Gloria Samuelson and Lottie Kearns.

agencies, but also to a lot of producers on America's farms and ranches.

Really, as we have said many times before, when you take into account the size of the total USDA budget, any cut in the budget really won't be noticed by any taxpayer because it is such a small part of the Federal budget.

To me, the most important and strongest lobbying effort should come from producers; you must be involved, and there is time to do it. Take the time to talk to your congressmen and senators as they come home to the district this summer to attend county and state fairs and to make their presence felt in the district. Make your presence felt too, by talking to them and offering your ideas on the ingredients important to you in the 2012 farm bill. If you can't talk to them personally, then let them know through the many other means of communication. Just remember to be polite and to the point. That could very well be one of the most important things you do for the future of your farm or ranch this year.

My thoughts on *Samuelson Sez.*

Secretary Vilsack Sez - December 31, 2011

For the past 35 years I have traveled to Washington, DC a few days before Christmas to sit down in the office of the Secretary of Agriculture, look back and look ahead. In addition to that, I've always invited the Secretary to share whatever he wants to with America's farmers and ranchers. That is why, once a year, we title this *Secretary Sez*, this year with Tom Vilsack.

"Let me, first of all, say "thank you" to the extraordinarily hard working farmers and ranchers of America. I have some sense of how hard you work, how difficult and risky your job is and your work is; but I also have an appreciation for how important it is. We often take what you do for granted in this country. You provide us with a security that very few countries and, people in very few countries have, of a food supply that is nutritious, that is diverse, that is available and affordable.

SECRETARY OF AGRICULTURE John R. Block (left) accepts a leather briefcase from Orion Samuelson, farm broadcaster from WGN/WGN-TV (Chicago), as a token of appreciation om the communicators who attended the first Agricultural Communicators' Congress.

In August, 1984 with Jack Block.

"You know, this year I have a slightly different view of the world. My nephew, Sam Bell, is a Marine and he's currently stationed in Afghanistan, in some remote area. I think of Sam often and pray for his safe return here to the United States. But I also think of the 55 people from USDA that are working with Sam Bell in Afghanistan; 55 individuals who are working with the farmers of Afghanistan to make sure that not only do they have a safe country, but they have a country with a working economy. They know that 80% of people in Afghanistan are connected to agriculture in one form or another. If we can get that agriculture in Afghanistan to be more profitable, then maybe we can make that country a bit safer, and then maybe Sam can come home, and then maybe we can be a bit safer.

"So you see, agriculture is not only about feeding our families, it is also about making this world a better and safer place. So, thank you very much for everything you do. God bless you and your families and I hope that 2012 is every bit as successful as you want it to be."

Well done, Secretary Vilsack, and as you look forward to the year 2012, a lot of travels on your itinerary, will you be representing American agriculture in other parts of the world?

"Likely so, but we will start with an opportunity in Hawaii at the American Farm Bureau convention. We are looking forward to the opportunity to talk a little bit about how we are dealing with reduced budgets, and how we are realigning and restructuring the U.S. Department of Agriculture to do a better job of providing services to farmers and ranchers and folks who work and live and raise their families in rural America."

And that is Tom Vilsack's thoughts on the *Secretary Sez.*

Lena passed away and at her graveside service the pastor had just said his final "Amen" and closed his Bible when there was a massive clap of thunder, followed by a tremendous bolt of lightning, accompanied by even more thunder rumbling in the distance.

Ole looked at the pastor and said, "Vell, she's dere."

Chapter Twenty-Three

And in the end...

Ole Samuelson couldn't have known what would lie ahead for himself, let alone his children and grandchildren, when he sailed from Norway to North America. Certainly, he couldn't have imagined what would happen to me. None of this... my career and accomplishments, nor this book, would have happened without the sacrifices of my ancestors and the support and love of family, friends, listeners, viewers and co-workers.

As I was celebrating my 50th year on WGN in 2010, it occurred to me that, my golly, I had been on the air for more than half of the life of that grand radio station! As for how much longer? Well, as I always say when I'm asked about retirement (and I'm asked often): I'm Norwegian, so I'm only half-done. But throughout my career, I've also said that when I'm not having fun anymore, it's time to quit. I came close to that a few years ago, when I dealt with management that didn't understand or appreciate the importance of serving the rural audience. Then a management shakeup brought in Tom Langmyer as WGN's General Manager and Bill White as Program Manager. These two gentlemen have become good friends who I trust and admire. They understand the importance of agriculture and the need to reach both urban and rural audiences with information. So... I am still here!

Back to the "R" word — which Gloria says causes me to stutter every time I try to say it — let me repeat what Paul Harvey told me about retirement: it is practice for being dead, and being dead doesn't require practice.

So maybe this chapter should be re-titled, "To Be Continued..."

If I am remembered past my days at the microphone, I hope it's for being an honest communicator who relied on common sense in presenting information that brought farm folks and city folks together. There are people in the city who are four, five and six generations removed from the farm and don't always understand what it takes to put food on the table. The two percent of our country's population that feeds us is comprised of the greatest people in the world and because of their loyalty to WGN, I've not had to go on food stamps!

I would also like to be remembered as a responsible communicator and guardian of the truth because I've seen over and over again how an uninformed and/or irresponsible media can erase years of farmers' and ranchers' hard work by sensationalizing non-stories, using a few misplaced, misused words, such as "swine flu," "bird flu," or "mad cow." You may recall how the bottoms fell out of the commodity and equity markets when those words were broadcast and printed in ways that scared consumers and investors. I hope to be remembered for caring more about getting it right than getting it first, and as a communicator who could express himself without using four-letter words. I'm that old-fashioned.

On a bookshelf in my office at WGN, I have a framed Cherokee prayer that serves as a reminder to me about communication:

"O Great Spirit,
Help me always
to speak the truth quietly,
to listen with an open mind
when others speak,
and to remember the peace
that may be found in silence."

One other thing for which I'd like to be remembered is as one of the best tellers of Ole and Lena stories, which I've scattered throughout this book. Here's one more of my favorites:

Ole, in his later years, spent a lot of time sitting on the front porch of 's farmhouse. As a tractor pulling a wagon drives by, Ole shouts, "Hey dere! re are ya headed?"

"I'm headed to my field," said the driver, who stops to chat.

"Vaht do you have in da wagon?" said Ole.

"Manure."

"Manure, eh? Vaht you gonna do vit it?"

"I spread it over my strawberries," said the farmer.

"Oh ya, sure," said Ole, "but ya should come over here for dessert sometime. Lena uses whipped cream."

Robin F. Pendergrast Photography, Inc.

Acknowledgements

One person does not make a book. It is the product of many people, and on these pages there are thousands of fingerprints from people who have touched my life over my long and ongoing career. Detailing my gratitude to them all could be a book in itself, and at the risk of offending someone who I inadvertently leave out, I will mention some of those whose fingerprints are most indelible.

First, my wife Gloria, who not only saved my life as you read in Chapter Five, but kept me on task, kept me organized, was my sounding board and proofreader and was my liaison to the publisher.

The skilled people at Bantry Bay Publishing of Chicago, primarily Steve Alexander and Diane Montiel, helped me finally make good on the promise I'd been making for many years: "It'll be in the book!"

The sharp eyes of proofreaders Lindsay Eanet and Sue Wolfe helped minimize typos, misspellings and grammatical errors that would have brought an "Uff dah!" from my high school English teacher, Mrs. Woods.

My gratitude also goes to the Abraham Lincoln Presidential Library and Illinois State Museum's Oral History of Illinois Agriculture Project and Mark DePue.

We received tremendous help collecting photos, including from WGN and the Tribune Company, Don Peasley, Brad LaPayne, Dick Resler, Robin Pendergrast, Max Armstrong, the USDA, and many others.

My sincerest thanks to the many people who've contributed their words to this book: Max Armstrong, Spike O'Dell, Roy Leonard, Lottie Kearns, Mike Johanns, John Block, Clayton Yeutter, Ann Veneman and Bob Stallman.

Finally, my appreciation for the love and support of my sister, Norma, my children, David and Kathryn, daughter-in-law Carla and grandchildren Matthew and Grace.